EXPLANATORY STRUCTURES

EXPLANATORY STRUCTURES

A Study of Concepts of Explanation in Early Physics and Philosophy

STEPHEN GAUKROGER

Research Fellow in the Philosophy of Science, Clare Hall, Cambridge

HUMANITIES PRESS · NEW JERSEY

This edition first published in 1978 by
HUMANITIES PRESS INC.,
Atlantic Highlands, NJ 07716

Copyright © Stephen Gaukroger, 1978

Library of Congress Cataloging in Publication Data

Gaukroger, Stephen.
  Explanatory structures.

  Originally presented as the author's thesis,
Cambridge University.
  Bibliography: p.
1. Physics—Philosophy. 2. Physics—History.
I. Title. II. Title: Explanation.
QC6.G285   1978   530′.01   78-14873

ISBN 0-391-00899-4

Printed in Britain by
Redwood Burn Limited, Trowbridge & Esher

CONTENTS

v

PART II

PREFACE

THIS book was developed from my Ph.D. dissertation for the University of Cambridge. It is concerned with the problem of how and why concepts of explanation change and, in particular, with the question of why concepts of physical explanation have been subject to profound changes in the development of physics. It is a central thesis of the present work that explanatory problems have to be treated in the context of the discourse in which they arise, since what counts as an explanation in a discourse depends on factors internal to that discourse which must be conceptualized and assessed as such. This project has required the presentation of detailed case studies which are designed to exhibit the constraints that particular discourses impose on what is required from an explanation, and hence to show why different kinds of explanatory problems arise in different discourses. The case studies are designed to offer philosophical elucidation of explanatory problems in a specific area — early physics — and they are not intended as original pieces of historical research. However, they do, I hope, clarify certain historiographical issues about how we are to conceptualize the relation between pre-classical and classical physics by focusing attention on the explanatory conditions under which a mathematical physics becomes possible: and, indeed, by showing just how important these explanatory conditions are.

The constant sympathetic advice of my Ph.D. supervisor, Mr G. Buchdahl, has been invaluable in the preparation of this work; as has that of Dr N. Jardine, who suggested numerous revisions to the penultimate draft. I am grateful to Professor M. B. Hesse, Professor G. E. L. Owen and Dr J. A. Schuster for their comments and advice on individual chapters; and to Ms B. A. Brown, Mr T. Counihan and Mr B. Hindess for their comments and advice on early drafts. Professor P. M. Rattansi and the editor of the book, John Mepham, made very helpful suggestions on preparing the work for publication. Finally, I am grateful to Ms L. Ferguson for checking the final draft and making some badly needed stylistic corrections.   S. G.

INTRODUCTION

CHAPTER 1

THEORETICAL DISCOURSES AND EXPLANATORY STRUCTURES

§1 Differentiation of Theoretical Discourses

THE questions that we shall be concerned with in this book can be stated quite simply. They are: How do we conceptualise the explanatory structure of a theoretical discourse? and, How do we assess the explanatory structures of particular theoretical discourses? I propose to deal with these questions at a general level, and then to examine two particular explanatory structures in detail. The purpose of this first chapter is to provide some preliminary clarifications.

To begin with, two terms need to be defined. A *theory* is anything which is, or can be, articulated in the form of a statement or set of statements which purport to offer, or which can be taken as offering, an explanation of something. A *theoretical discourse* is any unified set of articulated theories. The principles by which sets of theories are unified or differentiated are determined by the kind of analysis one is engaged in. There is an indefinite number of ways in which theoretical discourses can be classified: by chronological grouping, by subject matter, by formal characteristics, by authorship, and so on.

The first problem we must deal with arises from the fact that theoretical discourses do not inhabit a realm of their own. They exist in, and often persist throughout, particular social formations which can be characterised economically, politically, culturally and in a variety of other ways. At any particular period, complex relations hold between any one 'discipline' and another, and there are also complex relations which can hold between particular theories, mathematical and technological developments, and political, legal, educa-

3

tional and religious practices. This means that the origins and peculiar mode of development of a theoretical discourse may have a variety of determinants of quite different kinds. Indeed, there are, in principle, an indefinite number of factors which can have a bearing on the constitution of theoretical discourses. This raises a rather serious problem. We must determine whether a theory which can deal with all these factors is viable, or even possible, and if it is not whether any (non-trivial) account of theoretical discourses is possible which has a more restricted range but which is not based on an *arbitrary* decision as to the pertinence of particular factors.

The treatment of discourses as wholes requires that they be conceptualised in some way; at the very least, this means that we must be able to distinguish one theoretical discourse from another. The problems arise when we come to consider how we are to provide the basis for this differentiation. Theoretical discourses have commonly been differentiated on methodological grounds, on epistemological grounds, on the basis of straightforward dating, on the basis of the mode of organisation of the activities of scientific communities, and so on. It is clear that these bases of differentiation are not the same: to say that two discourses are epistemologically distinct is not the same as saying that they are methodologically distinct, or that the 'scientific communities' which produce them are sociologically distinct. Moreover, there is no reason, *prima facie*, to suppose that these bases of differentiation are extensionally equivalent. That is to say, methodologically distinct theoretical discourses cannot be assumed to be *eo ipso* epistemologically distinct, nor can the 'scientific communities' associated with these theoretical discourses be assumed to be *eo ipso* sociologically distinct. It is not too difficult to see how an account of theoretical discourses which lacks specific criteria for the differentiation of discourses risks simply running these various principles of differentiation together.

Consider, for example, Kuhn's account of the development of theoretical discourses.[1] He differentiates discourses in terms of *paradigms*. When we come to examine the basis of this differentiation we find ourselves in the realms of epistemology, law, psychology, optics and so on.[2] The problematic nature of differentiation in terms of paradigms is particularly transparent in Kuhn's account of the discontinuous transition between paradigms. This is conceptualised solely on the basis of one analogy: the Gestalt switch. However, the analogy is quickly rendered ineffective because Kuhn conceives paradigms in such a way that there can be a choice between them,[3] whereas the Gestalt switch is independent of voluntary motivations. Further, he prevaricates on the question of incommensurability, and in later essays he claims that he is not prepared to accept complete incommensurability,[4] which again ruins the analogy. This confusion results in an inability to characterise either paradigm change or the relations between paradigms. With regard to the former, we are told that a shift between paradigms 'must occur all at once (though not necessarily in an instant) or not at all'.[5] In making this statement Kuhn is presumably trying to accommodate the fact that 'scientific revolutions' take time to the theory that there is no continuous transition between 'world views'. That he cannot do this is clear from the unintelligible suggestion that something can occur all at once though not necessarily in an instant.

One of the main factors in the generation of problems of this kind is the wholly heterogeneous nature of paradigms. Now we might want to argue that there is, for example, an epistemological discontinuity between Aristotelian and classical mechanics, or between classical and quantum mechanics, as Koyré and Bachelard have done. Here we would be concerned with two systems of physics differentiated on the basis of incompatible accounts of what counts as knowledge and what access we have to reality. These epistemological issues are atemporal, and ahistorical, in a way that the development of scientific communities is not. Scientific communities, insofar as they consist of individual scientists practising at particular periods, are temporal phenomena. If we are concerned to give an account which is designed to link the relation between different communities and the relation between the epistemological structures of different discourses, this has to be borne in mind. Kuhn simply runs the two together in the concept of 'paradigm', and he is left with no basis on which to distinguish issues which make essential reference to time and ones which do not. The relation between the epistemological structures of two discourses, for example, may be discontinuous in that one is not a development of the other. In dealing with this issue we are concerned with questions which are quite different from those which would have to be raised in dealing with the chronological succession of discourses considered *in toto*: that is, considered as the products of particular social formations. Both these projects are worthy of serious consideration but they are different kinds of project and they involve different kinds of questions. Unless this is realised, no end of confusion will ensue.

In Kuhn's case, for instance, we are left with no idea of how a new paradigm is formulated; indeed, we do not even know its conditions of formulation, for although new paradigms are 'usually' formulated to deal with 'crises' in the old paradigm, they are not *necessarily* stimulated by crises.[6] Nor have we any clue as to the relation between the new paradigm and the one that it replaces, primarily because no information is forthcoming on the question of how the relation between paradigms is to be conceptualised. Indeed, since the concept of a paradigm provides us with no basis on which to relate — or even to *differentiate* — the various factors which play a role in the structure and development of theoretical discourses, it is highly unlikely that coherent and invariant principles of differentiation between theoretical discourses could be provided on the basis of an account of paradigms.

In rejecting Kuhn's project, I am not denying that the origins and peculiar mode of development of a theoretical discourse may have a variety of determinants of quite different kinds. Theoretical discourses are the products of highly complex cultural and educational practices, amongst other things. Nevertheless, this does not mean that everything which has a bearing on the constitution of theoretical discourses needs to be taken into account in the analysis of their explanatory structures, for example. It is true that insofar as various kinds of factor — social and otherwise — have a bearing on the constitution of theoretical discourses, they have a bearing on the constitution of the explanatory structures (or methodological structures or whatever) of those discourses. Nevertheless, an analysis of how an explanatory structure is constituted, and an analysis of that explanatory structure once constituted, are different kinds of project and involve different kinds of questions. The former may well involve reference to the social and cultural conditions under which a theory is produced; the latter does not. Our concern in this book is not with the constitution of explanatory structures as such but rather with the way in which explanatory structures, once constituted, function, and with the mechanisms by which explanatory problems are generated and resolved (if at all) in that structure.

I have argued that one of the most central problems in Kuhn's account of paradigms is the lack of any coherent criteria by which to differentiate paradigms. This problem arises because of a surfeit of candidates which are simply conflated; the result is that the criteria for differentiation of theoretical discourses which seem to be operative in Kuhn's work include too much. It may also be worthwhile, therefore, to consider a case in which the criteria appear to include too little. An example of such a case is, I think, Lakatos' methodology of 'scientific research programmes',[7] so it may be useful to examine Lakatos' work briefly in order to identify some of the problems that projects of this kind are prone to, so that we might be able to recognise and avoid such problems when we come to consider explanatory structures.

A *scientific research programme* is characterised by a 'hard core', a 'positive heuristic' and a 'negative heuristic'. The 'positive heuristic' consists of methodological rules which determine the paths of research to be pursued: it serves to forward the programme in terms of its sophistication and applicability. Those methodological rules which determine the research to be avoided are the 'negative heuristic', and they delimit that part of the programme which is kept free from falsification. It is the negative heuristic which keeps the hard core free from falsification by providing it with a 'protective belt' which directs all *modus tollens* falsifications to auxiliary hypotheses surrounding the hard core:

> 'It is this protective belt of auxiliary hypotheses which has to bear the brunt of tests and get adjusted and readjusted, or even completely replaced, to defend the thus-hardened core. A research programme is successful if it leads to a progressive problemshift, unsuccessful if it leads to a degenerating problemshift'.[8]

Now unless research programmes can be differentiated, we are not in a position to analyse and assess them. The criteria by which this differentiation is to be effected are obscure. First of all, we are told that 'even science as a whole can be regarded as a huge research programme with Popper's supreme heuristic rule: "devise conjectures which have more empirical content than their predecessors".'[9] However, what Lakatos is primarily interested in 'is not science as a whole, but rather *particular* research programmes, such as the one known as "Cartesian metaphysics".'[10] The word 'particular' as used here requires some clarification. To say that science as a whole can be regarded as a research programme is surely to say that it can be regarded as a particular research programme. What makes Cartesian metaphysics, or Newtonian dynamics, 'particular' cannot be their methodology in the general sense since this would not suffice to render them distinguishable from any other 'science'. Nor can it be their domain of investigation, since this would not serve to distinguish them from other physical theories.[11]

The most fundamental way in which Lakatos seems to want to differentiate research programmes is in terms of their 'hard core'. One way in which the hard core can be characterised is as whatever is kept free from falsification in a research programme. Hence, in discussing Newton's gravitational theory, Lakatos refers us to the fact that:

> 'In Newton's programme the negative heuristic bids us direct the *modus tollens* from Newton's three laws of dynamics and his law of gravitation. This "core" is "irrefutable" by the methodological decision of its protagonists: anomalies must lead to changes only in the "protective" belt of auxiliary, "observational" hypotheses and initial conditions'.[12]

Now if the hard core of Newton's gravitational theory is whatever is kept free from falsification in that theory, then I think it must include more than the four laws specified by

Lakatos. In the first place, derivations from these laws are
effected by the use of classical logic and mathematics, and
some philosophers have argued that classical logic, in
particular, is not irrefutable;[13] this may well lead one to
consider the necessity of including the laws of classical logic
in the hard core of Newtonian dynamics. Now the status of
classical logic may well still be contentious. There is however
a related issue which poses a similar (but much less contentious) problem about what to include in the hard core. This is
the fact that mathematical theorems can be used in the treatment of physical problems in Newtonian dynamics. This is
not the case in every physical theory: for example it is not the
case in Aristotelian dynamics. Indeed, this procedure is an
essential and 'irrefutable' part of Newtonian dynamics and I
cannot see that it is any different in status from irrefutable
laws. Moreover, the four laws are likely to be jettisoned
before this procedure is jettisoned in classical mechanics.
Similarly, it is far from clear that the postulate of the
independent existence of absolute (Euclidean) space and
linear asymmetric time is part of the 'auxiliary belt' of
Newtonian dynamics — it is much more likely that they are
part of the hard core.

These considerations are neither fanciful nor purely
hypothetical, since they have a crucial bearing on the question of assessment. For Lakatos, a problemshift is 'theoretic-
ally progressive' if it represents an increase in empirical
content, and it is 'empirically progressive' if this increase in
content is corroborated. Further, a theory 'is "acceptable"
or "scientific" only if it has corroborated excess empirical
content over its predecessor (or rival), that is, only if it leads
to the discovery of novel facts'.[14]

The question of excess empirical content is closely related
to what we take the hard core of a research programme to be.
Feyerabend,[15] for example, has questioned the content of the
hard core in the two main Lakatosian accounts to date: the
Zahar account of the Lorentzian and Einsteinian research
programmes to 1915,[16] and the Lakatos-Zahar account of
the Ptolemaic and Copernican systems.[17] The problems arise
here not simply because it is an open question what one includes in the hard core of a rationally reconstructed research

programme, but also because the decision about what is to count as the hard core can completely alter one's assessment of a research programme. In his discussion of the Ptolemaic/Copernican and Lorentz/Einstein programmes, Feyerabend manages to reverse the conclusions of Lakatos and Zahar by adding an Aristotelian dynamics and cosmology to Lakatos' characterisation of Ptolemy's research programme, and by adding atomism to Zahar's characterisation of Lorentz's research programme. These additions enable Feyerabend to argue that the Ptolemaic and Lorentzian programmes have greater empirical content than their rivals and that they are more progressive research programmes in Lakatosian terms. No-one could deny that the Ptolemaic system, for example, includes (or is supportable in terms of) a particular cosmology and dynamics, but because of Lakatos' failure to specify how we determine hard cores in a sufficiently consistent and comprehensive manner, we do not know whether these should be included in the hard core or not, and this failure renders the project of assessment impossible.

The fundamental problem with Lakatos' project is, then, that there is a discrepancy between the general criteria which he presents for distinguishing 'hard cores' (*viz*, whatever is kept free from falsification), and the more restrictive criteria that he appears to operate with in practice: criteria which result, in the case of physics for example, in only the primitive physical laws being taken as the hard core. This latter procedure is rather arbitrary, and it is far from clear that primitive physical laws provide sufficiently comprehensive criteria for differentiating research programmes in physics. Moreover, one of the central features of Newtonian physics is that it is a mathematical physics — unlike Aristotelian physics (or Cartesian physics for that matter) — and this is something which Lakatos cannot deal with if the primitive physical laws alone are taken as the hard core. As I hope to show in later chapters, the use of mathematics in physics imposes important constraints on the ways in which physical problems can be posed and resolved, and these constraints are quite different from those which are operative in the case of Aristotelian physics, for example. Constraints on posing and resolving problems clearly have an important bearing on the ways in which physical laws are developed and assessed: Lakatos' project can offer no elucidation on this issue and, to this extent, its scope is rather limited.

With regard to the first criteria (which specify that anything which is kept free from falsification be included in the hard core), since these have never been applied in any case studies it is rather difficult to assess their usefulness. It may be noted, however, that Feyerabend's additions to the hard cores of the research programmes that Lakatos and Zahar have studied in detail do have clearly relativistic consequences, in that the Ptolemaic and Lorentzian programmes do actually come out as being more 'progressive' than their successors. In view of this, we could only expect the further additions which would be required, if everything that was to be kept free from falsification was included in the hard core, to result in even more relativistic consequences. This would obviously render the assessment of research programmes highly problematic.

There are three general points that can be made at this stage with regard to the question of differentiation of theoretical discourses. First, theoretical discourses are not differentiated 'in nature'. They can only be differentiated on the basis of some theoretical classification. It is true that there are some very prevalent classifications — such as that of the 'sciences' into the 'natural', 'social' and 'mathematical' sciences — which it may seem very difficult to do without. But different classifications — which may cut across these — are *possible*, and the classification mentioned is implicitly theoretical. Moreover, its heuristic success should not lead us to think that it is successful for all purposes. Secondly, theoretical discourses cannot be differentiated as such, they can only be differentiated in virtue of their having distinct epistemological structures, or explanatory structures, or research programmes or whatever. Correlatively, it is uninformative to speak of theoretical discourses as being commensurable or incommensurable *simpliciter*, because there are an infinite number of respects in which discourses can be commensurable or incommensurable. Thirdly, different characterisations of discourses may include very different kinds of factor, depending on the questions that we want answers to. Some questions, for example, will not require reference to the social conditions under which a discourse is produced and developed, whereas others will not require reference to the epistemological structure of discourses. What factors require consideration when dealing with specific issues is not something that can be decided by fiat, and theoretical innovations which add to our understanding of the development of discourses have often occurred as a result of arguments showing that problems posed by earlier accounts could not be answered properly in the terms in which they had been set up.

In connection with these three general points, it may be remarked that the way I have set up the problem of differentiation of theoretical discourses precludes any *informative* distinction being made between 'internal' and 'external' problems. How one differentiates discourses is certainly 'conventional' — where these 'conventional' differentiations may be assessed, as we have just seen — but it is not arbitrary since it depends on what questions one wishes to pose, and hence on the factors which are going to be taken into account. The 'internal'/'external' distinction presupposes that there is some central core of theory and sets of extraneous factors which may or may not be pertinent to the treatment of certain questions concerning this hard core. I can see no justification for such a presupposition. In analysing Galilean astronomy in purely methodological terms — in determining whether its methodology is coherent and viable — a specific range of questions can be posed which require the introduction of specific kinds of factors. In analysing Galilean astronomy in terms of its epistemological structure — in terms of its account of reality and its account of what access we have to reality — quite different questions will be posed and a different range of factors taken into consideration. Similarly, in analysing the problem of why the Copernican theory was not generally accepted in the late sixteenth and early seventeenth centuries different questions will again have to be posed and factors such as Catholic dogma and the role of the Church would have to be taken into consideration. Which questions we want answers to will determine how we characterise discourses and what factors we include in this characterisation. We cannot simply expect sociological accounts of discourses to illuminate questions which are more properly of an epistemological nature and vice versa; not can we expect purely methodological accounts of discourses to illuminate questions of a more 'metaphysical' nature.[18] And it is, indeed, for reasons of this kind that the currently popular attempts to see discourses as social, political and theoretical totalities are of no analytical use whatsoever.

In analysing the explanatory structures of two physical discourses in Part II, I shall not be attempting to illuminate *all* kinds of conceptual problem which arise in a discourse by referring to its explanatory structure, nor shall I be attempting to *reduce* the discourses in question to their explanatory structures. I shall be concerned with a definite range of problems. While this range of problems, which is circumscribed by the kind of analytical concepts that I shall introduce, is limited, the problems are important ones if we wish to understand how explanatory failures are generated and how they are to be overcome.

§2 Explanation

Up to now, I have argued that there is an indefinite number of ways in which theoretical discourses can be differentiated; I have also tried to show how problems of differentiation can bear on the questions of analysis and assessment. The primary aim of the first part of our project is to provide criteria by which discourses can be differentiated in terms of their explanatory structures, in such a way that these explanatory structures can be analysed and assessed.

Theories — and theoretical discourses — have been defined in terms of 'explanation', and it has been stipulated that a necessary condition for something's being a theory is that it explains or purports to explain something. To explain something is to render it intelligible, and we shall be concerned with theories only insofar as they are designed to render things intelligible.

The idea of rendering something intelligible requires elaboration, and there are several ways in which this can be done. The kind of account of explanation that we are seeking is not one which is designed to provide a logic of explanation, nor is it one which attempts to specify what we would currently count as a good explanation. This being the case, we shall not be concerned with developing and assessing the accounts which are directed towards these questions; in particular, we shall not be concerned with the relative merits of the deductive-nomological and statistical-relevance models of explanation. Rather than attempting to determine what is a good explanation *per se*, or trying to provide an account of the necessary and sufficient conditions for explanation in 'science', the aim of the project in which we shall be engaged is to provide the means whereby we can isolate and analyse what counts as an explanation in different discourses. The *explanatory structure* of a theoretical discourse can be defined simply as that structure which determines what counts as an explanation in the discourse. We shall be concerned to discover how explanatory structures are composed, and in doing this we shall be considering particular explanations *solely* as products of particular explanatory structures, so that when it comes to the question of assessment no attempt will be made to assess explanations independently of the discourse in which they are produced.

The factors which play a part in determining what counts as an explanation in a particular discourse are not easy to specify, but some general features can be discerned. An explanatory structure operates with a set of entities in terms of which explanations are given. The ultimate irreducible set of such entities is the *ontology* of the discourse. In explanation, an ontology is linked to a *domain of evidence*, inasmuch as it is a necessary condition of a statement's being a candidate for an explanation that there be specifiable circumstances in which the claim of that statement would be unsound. The domain of evidence of a theoretical discourse is the set of those phenomena which could confirm, establish or refute purported explanations in a particular discourse. *Phenomena* is used here (provisionally) in the widest sense, and it can include anything from Papal fiats to planetary motions, depending on the discourse in question. Insofar as it operates with a domain of evidence, an explanatory structure operates with criteria for what does and what does not count as evidence in a particular discourse. These criteria are criteria for what could count as evidence, not for what the evidence is: what the evidence is can only be established by the procedures of investigation with which that discourse operates.

The link between the ontology of a discourse and its evidential domain consists of a *system of concepts* peculiar to the discourse, and a *proof structure* which circumscribes the class of valid and invalid consequences and derivation relations which may hold between any statements in the discourse. In determining what counts as a proof in that discourse, it also determines whether concepts, techniques, etc. which are products of different discourses need to be introduced in order that explanations might be generated. For example, in the case of some physical discourses the ontology and domain of evidence are connected by a set of physical concepts, but the kinds of proof or demonstration which are required in these discourses necessitate the introduction of mathematical concepts and techniques into the proof of physical theorems.

In short, an explanatory structure consists of an ontology, a domain of evidence, a system of concepts relating these two, and a proof structure which specifies the valid relations which can hold between the concepts of this system. The idea of an explanatory structure will be discussed in detail in subsequent chapters. For the moment, it is sufficient to note that I am proposing that we differentiate theoretical discourses in terms of their explanatory structures: hence the importance of defining theories as statements, or sets of statements, which purport to explain. I now want to turn to the question of what exactly it is that theories and theoretical discourses explain.

As a first approximation, we may say that theoretical discourses explain those phenomena which fall within their domains of investigation. That this *is* an approximation is something that I shall try to establish later. The immediate problem is why particular phenomena fall within the domains of investigation of particular theoretical discourses. In dealing with this question we must be able to give some account of how the domains of investigation of theoretical discourses are constituted. I use the word *constituted* advisedly here. That the domains of investigation of discourses are not 'given in advance', so to speak, is clear from the fact that the domain of investigation of any discourse is subject to development as the discourse itself develops. One need think only of the inclusion of electro-magnetic phenomena in the domain of investigation of classical dynamics in the early nineteenth century, or the inclusion of health — as well as disease — in the domain of investigation of medicine in the same period.

Now the domain of investigation of a discourse is demarcated by the concepts of that discourse, and these concepts specify certain ranges of phenomena. I want to look first at the reference or extension of domains of investigation, and particularly at the problem of how we decide whether there is, or can be, an overlap between the domains of investigation of different discourses. I shall then consider the more complex issue of how the domain of investigation of a discourse can be specified in terms of state-variables.

§3   The Reference of Domains of Investigation

One of the most straightforward necessary conditions for the comparative assessment of two explanations which are produced in different discourses is that the two explanations should be explanations of the same thing: one explanation may be more general than the other (in that it may explain more than the other), but this is not a problem as long as there is an overlap in what is being explained. It is, for example, usually considered inappropriate to compare the explanations of physics and economics because there is no overlap between the domains of investigation of these disciplines. The explanations of physics and economics do not compete because the same kinds of things are not being explained. Now a *discipline* can be characterised in terms of its domain of investigation. Two theories having the same domain of investigation are within the same discipline. We speak of Aristotelian physics, Cartesian physics and

Newtonian physics, for example, as belonging to the same discipline. This has an intuitive plausibility, and the idea of a discipline is often adequate for everyday purposes. The problems arise when we try to specify, in more precise terms, what it is that makes Aristotelian 'physics' and Newtonian 'physics' part of the same discipline. In calling both these theories *physical* theories we presuppose that there is some fundamental connection between them, and one such connection that has generally been regarded as important is 'stability of reference': that is to say, it has been considered crucial that we be able to establish that Aristotelian 'physics' and Newtonian 'physics' are, at least partially, about the same kinds of thing.

Before considering how this has been attempted, I think it is worth distinguishing two different areas of reference which a theory has. A theory can be said to refer to those entities or phenomena which are being explained, but it also refers to those entities which it invokes to explain what is being explained. For example, I may invoke atoms to explain the properties of a substance such as gold. Atoms would figure here in the ontology of the theory, and gold would figure in the domain of investigation of the theory. Similarly, if I am explaining the properties of atoms in terms of subatomic particles then atoms figure in the domain of investigation and subatomic particles in the ontology. The conceptual distinction between a domain of investigation and an ontology is an important one. It is one thing to attempt to establish that there is stability of reference between the domains of investigation of a Greek atomist account of *chrusos* (which we translate as 'gold') and a modern chemical study of gold, for example, but it is quite another to attempt to establish the much more contentious thesis that in the Greek atomist account of gold in terms of *atomoi* (which we translate as 'atoms'), and in a modern chemical account in terms of atoms, there is stability of reference between '*atomos*' and 'atom'.

Claims about stability of reference between domains of investigation become problematic when we consider the ways in which domains of investigation are conceptualised. For instance, if we take the domain of investigation of physics to be changes in energy distribution or, alternatively, if we take it to be motion, then it depends on how we construe these terms whether we include what appears in Aristotle's *Physics* under the title of 'physics' at all. It would now be considered an anachronism to equate (at the level of meaning or 'sense') *energeia* with the modern concept of energy, or to equate *metabolē* or *kinēsis* with the classical concept of motion. The Aristotelian concepts have demonstrably different senses from the modern ones. The problems here arise not because one domain of investigation is conceptualised in two different ways, but because of the possibility that two referentially distinct domains of investigation may be involved. This possibility arises because of the relation between sense and reference on the traditional Fregean account of meaning. One way in which this doctrine can be expressed is to say that the reference of a term is determined by its sense, inasmuch as its reference consists of that set of individuals which satisfy a set of conditions which must be fulfilled if the term is to be used with its correct sense. In the limiting case (such as in well-developed physical theories) these may be necessary and sufficient conditions: in other cases there may be a considerable degree of ambiguity.

A crucial point to note about the sense/reference distinction is that if we characterise domains of investigation in terms of classes of individuals which satisfy the specifications of the set of concepts which are used to demarcate those domains of investigation, a difference in these sets of concepts does not, *in itself*, entail that there is no individual which is a member of both domains. Hesse, for example, has noted that there is a considerable overlap between the extensions of the classical and Relativistic concepts of mass insofar as there is a large number of situations which can be described in a logically comparable way in both classical and Relativistic mechanics.[19] Now although there is nothing exceptional about this example, it is not representative of all the cases in which we may want to relate two domains of investigation. It is a relatively unproblematic example because there are well-known procedures by which classical mechanics can be treated as a limiting case of Relativistic mechanics. But no such treatment is possible in the case of 'motion' as this figures in Aristotelian and classical mechanics, for example, and comparison here is much more problematic. This is not to say that, other things being equal, such a comparison is *impossible*. On the Aristotelian account of 'motion' (*kinēsis*), 'motion' is treated as a species of change in general. That is, in giving an account of 'motion' this account must be related — in terms of the way in which it is conceptualised — to such phenomena as generation and corruption, growth and so on. But this, *in itself*, does not mean that the Aristotelian account of motion is an account of something which is (referentially) quite different from what we now call motion. Not does the fact that 'local motion' is construed in terms of change of place (*topos*) — as opposed to change in position in space — have any such implications. These two considerations *in themselves* would not preclude there being situations in which there is an overlap in reference.

Up to now, I have argued that the sense/reference distinction does not preclude the possibility of there being domains of investigation which are conceptualised in different ways but which nevertheless have some part of their extension or reference in common. What it does *not* do is to guarantee stability of reference across domains of investigation which are conceptualised in quite different ways: whether two domains of investigation actually have any shared reference is something which is an *a posteriori* matter, and the only way it can be decided is by an examination of the discourses whose domains of investigation these are.

Obvious as this point may seem, it has in fact been denied by some philosophers — notably Putnam,[20] who has proposed an account of reference which makes stability of reference an *a priori* matter. Putnam's line of reasoning is best illustrated by an example. Let us say that Archimedes' criterion for something's being *chrusos* is that it have a specific gravity of 5 units. Our criterion for something's being gold is that it have 79 protons per atom. *Chrusos* and gold, in Fregean terms, have different senses. Also in Fregean terms, they come out as having two different references in some cases since a perfect 1:1 alloy of two metals which have specific weights of 4 and 6 units respectively, comes out as *chrusos* on Archimedes' criterion, but it would not come out as gold on our criterion. On Putnam's account, on the other hand, *chrusos* and gold *do* come out as having the same extension or reference, and Archimedes simply did not have the correct criterion for something's being gold:

> 'In the view I am advocating, when Archimedes asserted that something was gold (*chrusos*) he was not just saying that it had superficial characteristics of gold (in exceptional cases, something may belong to a natural kind [gold is a "natural kind" for Putnam] and *not* have the superficial characteristics of a member of that natural kind in fact); he was saying that it had the same general *hidden structure* (the same "essence", so to speak) as any normal piece of gold. Archimedes would have said that our hypothetical piece of metal X [in our example this is the 1:1 alloy] was gold, but he would have been *wrong*. But *who's to say* he would have been wrong? The obvious answer is: *we are* (using the best theory available today).'[21]

In short, Archimedes believed that all instances of *chrusos* have a common structure. Whereas he did not know what this common structure is, however, we do: it is atomic structure.

Putnam's claim is that gold has an essential nature and (as far as we know) this is represented in its atomic structure. That is to say, gold is *essentially* element 79. Because of this, when we say that *chrusos* and gold refer to the same thing, the 'sameness' in question is, and (as far as we know) can only be, atomic structure.

The trouble with this argument is that whereas there may be, in principle, evidence which would enable us to determine whether any statement were universally true, there is nothing that could count as evidence for a statement's being necessarily true. There is no way, for example, by which we can tell which sentence of the form '$\forall x[(x \text{ is gold}) \text{ iff } (x \text{ is } F)]$' is the one which expresses the necessary identity and hence provides a statement of the essential nature of gold. Now our not being able to tell whether or not particular statements are in fact essential would not preclude there being essential statements: just as our not knowing which of a set of statements is true would not prevent one of them being true. But if — as appears to be the case — there is no way, in principle, of telling what could possibly count as evidence for any statement's being essential then the idea that atomic accounts of gold express the 'essential nature' of gold becomes vacuous.[22]

In short, I cannot see that the kind of account that Putnam offers has any advantages over a Fregean account of sense and reference. The strength of the Fregean account is that it allows us to establish stability of reference on *a posteriori* grounds, and not on the grounds of some very contentious *a priori* guarantee. This has two great advantages. First, it does not allow us to establish stability of reference where we would like it to occur but only where it actually does occur. If these latter cases are fewer than we had previously thought then we have learned something which we could not have learned had we followed Putnam's Whiggish recommendations. Secondly, it allows us a much greater degree of sophistication in assessing explanations: we are not simply restricted to 'true' and 'false'. I shall consider the problem of assessment in the final section of this chapter but it is perhaps worth noting here that, in establishing that there is a referential overlap in the domains of investigation of two discourses, we have certainly fulfilled a necessary condition for the comparative assessment of individual explanations produced in the two discourses, but we may well not have satisfied the sufficient conditions for this. It is equally important that we be able to establish that what counts as an explanation is the same in the two cases.

This second condition may appear to restrict the range of assessable cases considerably, especially when compared to those accounts which simply stipulate that there should be an overlap in reference. But it is usually an implicit assumption of such accounts that variations in what counts as an explanation do not arise. Indeed, the kind of account which Putnam offers precludes such a variation, since for Putnam there seems to be only one way of explaining things. This takes the form of statements about 'essential natures', and attempts to explain things (at least in such areas as physical and chemical theory) can always be construed in this fashion. All these claims are in fact explicit in the Archimedes example in the passage from Putnam that I quoted above, and it is implicit in the passage that the example is generalisable. On a Fregean account of sense and reference, on the other hand, a variation in what counts as an explanation is not precluded — since Frege's account is quite 'neutral' with respect to this issue — but assessment of individual explanations is restricted to cases in which it does not occur. The reason for this is straightforward. In cases where a difference of this kind does arise it is not sufficient to establish stability of reference, since this in itself would not enable us to carry out a comparative assessment of two different explanations (or purported explanations) since the explanations would have different aims. Our main concern will, in fact, be with the cases in which there is a radical variation in what counts as an explanation. In particular, we shall be examining Aristotelian and classical mechanics and, I shall argue, what counts as an explanation in these discourses is quite different. Because of this, we shall not be able to carry out any comparative assessment of individual explanations from Aristotelian and classical mechanics.

§4   The Constitution of Domains of Investigation: State-Variables

Up to now, I have concentrated on the reference of domains of investigation. This has meant treating domains of investigation in terms of sets of phenomena. If taken in isolation, this kind of account is liable to be misleading. Domains of investigation have reference in virtue of the fact that they are characterised by sets of concepts which have senses. All I have tried to establish up to this point is that concepts with different senses may have the same reference, and because of this the domains of investigation of two discourses may overlap referentially. We must be very careful here however. Take, for example, the hypothetical case of two different discourses whose domains of investigation have identical reference. This is *not* a case of two discourses sharing the one domain of investigation. In virtue of the fact that we have two different discourses, the domains of investigation will be conceptualised in different ways, different problems will be posed and different kinds of solutions will be sought to these problems. As well as in referential terms, domains of investigation can also be construed as specific sets of problems requiring specific kinds of treatment. Logically, the conceptual specification of a domain of investigation is prior to its referential specification. Aristotelian physics, Galilean physics, Cartesian physics and Newtonian physics are all *physics* because the concepts in terms of which their domains of investigation are conceptualised have overlapping reference. They are not all *physics* because there is some 'pre-theoretical given' which consists of 'physical phenomena', and which different theories conceptualise in different ways. 'Physical phenomena' *is itself a concept* — or, more strictly speaking, a series of different concepts depending on the discourse that we are operating within — *so it cannot be conceptualised.*

I now want to turn to the treatment of domains of investigation as sets of concepts in terms of which specific problems can be posed. The first question arises from the fact that I have characterised theoretical discourses in explanatory terms, as unified sets of theories which purport to explain certain ranges of phenomena. This might seem to suggest that the domain of investigation of a discourse is simply that range of phenomena which require explanation in that discourse. Things are not quite as simple as this however. An an illustration of one set of factors which serves to complicate the issue let us take those parts of the domains of investigation of Aristotelian and Newtonian dynamics that deal with 'local motion' (or, in Aristotelian terms, '*kinēsis* in respect of place'). We can construe these domains of investigation in terms of systems.

As a preliminary definition — which holds for the present example but not generally outside physics — we can say that a system (which, in the limiting case, may be an isolated body) is in a *state* if it is in a situation which does not require a force for its maintenance; it is in, or is undergoing, a *process* if it is in a situation which does require a force for its maintenance. More generally, systems can be characterised in terms of state-variables; where the values of these variables are constant the system is in a state, where the value of one of the variables changes the system is undergoing a process. In Aristotelian dynamics, for example, rest is a state and motion is a process; in Newtonian dynamics, rest and uniform rectilinear motion are states and all other kinds of motion are processes. The difference here can be put down to the fact that place is a state-variable in Aristotelian dynamics whereas in Newtonian dynamics place (or position in space) is not a state-variable.

We can see from this illustration that state-variables are extremely important with respect to explanation since they serve to demarcate those situations which require explanation from those which do not. In Aristotelian dynamics, for example, uniform rectilinear motion requires dynamical explanation; in Newtonian dynamics it does not. Moreover, to say that uniform rectilinear motion does not 'require' explanation in Newtonian dynamics is not to say that it can be explained but that there is no need to explain it; rather, it means that it is not a proper object of explanation in that discourse.

Now the decline of the Roman empire is also not a proper object of explanation in Newtonian dynamics (unless, of course, one is a Laplacian reductionist). But this does not confer on it the same status as uniform rectilinear motion *vis-à-vis* classical dynamics. The distinction between the two cases is that between those phenomena which fall outside the domain of investigation and those which fall inside it. The decline of the Roman empire falls outside the domain of investigation of Newtonian dynamics for the simple reason that it cannot be characterised in terms of the state-variables of Newtonian dynamics. Uniform rectilinear motion falls inside this domain of investigation because it *can* be characterised in this way. At a general level, we can define (the reference of) the domain of investigation of a discourse as all and only those phenomena that can be characterised in terms of the state-variables of that discourse.

Now because uniform rectilinear motion involves no change in the values of state-variables, it does not require explanation in Newtonian dynamics. We can express this fact by saying that, although it falls within the domain of investigation of Newtonian dynamics, it falls outside the *domain of explanation* of this discourse. The domain of explanation of a discourse is always a subset (indeed, I suspect that it is always a *proper* subset) of the domain of investigation; it can be defined as the subset of the domain of investigation which includes all and only those phenomena which involve a change in the values of state-variables.

The choice of state-variables will thus determine not only what phenomena figure in the domain of investigation but also which of these phenomena require explanation. *That such a choice cannot be arbitrary is clear from these two facts*. The problem is: how is this choice controlled? In answering this question we must distinguish state-variables from the procedures by which state-variables are established. A variation in state-variables is not the same as a variation in the procedures by which state-variables are established. It is of considerable importance that we be able to distinguish between changes of state-variables which are due to changes in the procedures by which state-variables are established, and changes in state-variables which are not due to this.

An example of the second kind of case is, I think, Ellis' reformulation of Newton's first law of motion, which reads as follows:

> 'Every body has a component of relative acceleration towards every other body in the universe directly proportional to the sum of their masses and inversely proportional to the square of the distance between them — unless it is acted upon by a force.'[23]

This law is derivable from the law of universal gravitation and the second law of motion. If it were adopted, revisions would be required in our ideas about such things as stress and momentum, and we would have to equate gravitational and inertial mass, but the law nevertheless holds in any world in which Newtonian dynamics holds, and it does not hold in any world in which Newtonian dynamics does not hold.

The peculiar feature of the reformulated law is that rest is no longer a state in all situations, since a body can be at rest *and* be subject to the action of unbalanced forces. This means that whereas rest falls outside the domain of explanation of Aristotelian and Newtonian dynamics, in some circumstances it falls inside the domain of explanation of Ellis' reformulation. There is an apparent similarity between these three cases in that they all seem to concern variations of the domain of explanation within the one domain of investigation. My argument will be that this similarity is more apparent than real.

Most problematic in this respect is the relation between the state-variables of Aristotelian and classical mechanics. In a discussion of this topic, Cummins has argued that there are discrepancies in the Aristotelian account of motion which do not occur in classical mechanics, and that these discrepancies arise from the state-variables which are chosen — particularly from the choice of 'place' as a state-variable. Cummins argues that the following set of formal conditions for state-variables hold in both Aristotelian and classical mechanics:

> 'Given a system s, we may not choose a set of state-variables such that there is some interval during which only one variable changes. If this condition is not met, then one of three things must hold: (1) the system is not closed, (2) the set of state-variables is not complete, i.e. there is a "hidden variable", or (3) the changing variable is spurious, i.e. changes in the value of the variable do not represent changes in the sort of state under consideration.'[24]

He then puts forward an imaginary experiment in which, if this set of conditions is not to be violated, anomalies arise in the Aristotelian account of the experiment but not in the classical account.

We are asked to imagine two situations. In the first, we have an inclined plane up which a body is being hoisted with a uniform velocity. Since uniform velocity is a state the force (F_1) with which the body is maintained in uniform rectilinear motion must be equal and opposite to the force (F_2) acting to move the body downwards. On the Peripatetic account, however, uniform rectilinear motion is a process, so F_1 must be greater than F_2. Let us say that, on this latter account, $F_1 = F_2 + f$. We can also assume, along with Cummins, that the component forces of F_2 — gravitational force, air resistance and surface forces — can be treated as constant.

Now f must be a constant force since, on the Peripatetic account, it is responsible for uniform motion: when the body is at rest $F_1 = F_2$, when it is moving up the plane with a uniform velocity $F_1 = F_2 + f$. Cummins now suggests a second situation in which 'a ball of the same mass [is] placed on the plane inclined at angle A from a plane normal to g such that the resultant of g and the force exerted on the ball by the surface [i.e. F_2] equals f.' He then presents the 'empirical refutation':

> 'According to the Peripatetic theory, this ball should move with a uniform velocity v down the plane. But in fact, for any detectable angle, and reasonably large and heavy bodies, there will be a constant acceleration right through v (which is fairly small in this case)'.[25]

The basic problem is, then, that the same force seems to be responsible for both a uniform and an accelerated motion since, on the Peripatetic account, a constant force is responsible for a uniform motion when the body is moving up the inclined plane, but it is also responsible for an accelerated motion when the body is moving down the plane.

Now this is certainly a serious anomaly in Peripatetic mechanics, although it is one which later Peripatetics were well aware of, and *impetus* theory was designed to deal with a problem exactly analogous to this — the problem of how a constant force ('gravity') could produce the known acceleration of falling bodies. One might conclude from Cummins' reasoning that the common ground which he attempts to provide between Aristotelian and classical mechanics is sufficient to allow dropping place as a state-variable from Aristotelian mechanics. Such a conclusion would, I think, be seriously wrong. This is not to deny that there are fundamental problems in Aristotelian mechanics concerning the characterisation of states and processes. The crucial point, however, is that classical mechanics does not so much resolve these anomalies as provide a different system within which they do not arise. Cummins' argument often appears to suppose that the range of state-variables from which a choice can be made is common to Aristotelian and classical mechanics. Not only is this supposition false, but it would also be wrong to suppose that the procedures by which state-variables are established, and the criteria by which their appropriateness is decided, are exactly the same for the two discourses. It is to a discussion of this issue that I now want to turn, and I want to establish that it is ultimately misleading to construe Aristotelian, Newtonian and Ellis' dynamics in terms of variations of domains of explanation within the one domain of investigation, for strictly speaking there are two domains of investigation involved: one of these is that of Aristotelian dynamics, while the other is that of classical (Newtonian and Ellis') dynamics.

In characterising Aristotelian and Newtonian physics at the beginning of this section, I said that we would consider those parts of the domains of investigation of these discourses which deal with local motion. For many purposes there may be no need to introduce the other areas covered, but in the present context it is vital that we take them into account, not for their own sake but because of the light they may shed on the way in which state-variables are established at a general level. Local motion is not the only phenomenon in the domain of investigation of Aristotelian physics. Moreover, since neither Aristotle nor the later Peripatetic physicists are engaged in a reductionist programme — unlike the Atomists, for example, they do not wish to reduce all the phenomena of change to the shapes and motions of atoms — local motion has no priority of any kind over the other phenomena that figure in this domain.

For Aristotle, the domain of investigation of physics concerns those things which are subject to change. There are two different ways in which change can occur on this account: the subject in which properties inhere can itself be changed, or one of the properties can be changed. Given this distinction, there are several kinds of change which are possible. These kinds of change are limited by Aristotle's list of 'categories', which is the list of the widest kinds of predicates which are predicable essentially of nameable entities. These categories include such things as substance, quantity, quality, relation, place, date, action and passivity. For Aristotle, explanations take the form of definitions of the essences of phenomena, and the function of the categories is to provide a classificatory schema which enables us to differentiate kinds of entity at a preliminary level in terms of the kinds of predicate which are applicable to them. Local motion is construed in terms of the category of place, and changes of place are subject to the same general kind of treatment as changes within any other category. Place appears as a state-variable because of the doctrine of categories and this doctrine is central to the explanatory structure of Aristotelian physics: which means, in turn, that the rationale behind the postulation of place as a state-variable is something which can only be examined in the context of the kind of project in which Aristotle is engaged.

I shall discuss this project in detail in Part II, and I hope to show there that it is radically different from Galileo's and, by extension, from that of Newtonian dynamics. It is because these projects are different — because they operate within discourses with different explanatory structures — that different constraints are imposed on the procedures by which state-variables are established. This has an important bearing on how we conceive of Ellis' first law of motion. Ellis' reformulation is a revision made explicitly within Newtonian dynamics. This means that it is subject to the same constraints with regard to the question of the procedures by which state-variables are established; indeed, when Cummins argues (with qualifications) that the reformulated law constitutes an improvement on Newton's first law he implicitly assumes this, and in this case the assumption is a correct one. But it is because these constraints differ from those which are operative in Aristotelian dynamics that we can distinguish between Aristotelian dynamics, on the one hand, and Newton's and Ellis' dynamics on the other; and it is because of this distinction that Newton's and Ellis' choice of state-variables can be subjected to a direct comparison whereas in the case of Aristotelian dynamics this is much more problematic. The comparative problems arise in this latter case because of the dependence of the procedures for establishing state-variables on the explanatory structure of the discourse concerned.

§5 Assessment of Explanations and Assessment of Explanatory Structures

To conclude this introductory chapter, I want to make some general and schematic points about assessment. Before a comparative assessment of two explanations can be made, two things must be established. First, what is being explained must be the same in the two cases. Secondly, what counts as an explanation must be the same. Cases which fulfil both these requirements are relatively straightforward, and evaluation of the two explanations can be carried out in terms of such things as degrees of evidential support, generality, accord with other well-established theories, and so on. There is no shortage of literature on these very interesting topics.

The cases that we shall be concerned with primarily are those which fail to fulfil the second requirement. In such cases, the area of conflict (if it be such) lies not simply in the explanations themselves and their respective support and generality, but in the more fundamental problem of what should be sought in explaining certain kinds of phenomena. What is at stake in these cases is *what counts as an explanation*, and this is an issue which requires an analysis and assessment of the explanatory structures involved.

We cannot carry out a comparative assessment of individual explanations from discourses with different explanatory structures. This does not mean that we cannot establish some stability of reference between Aristotelian and Newtonian accounts of motion, for example, and show that one of these produces anomalies in its own terms, whereas the other does not. We have, of course, just done this. In the case we considered, the anomalies lay with the Aristotelian account, but we could just as easily have demonstrated anomalies in Newtonian mechanics which do not arise in Aristotelian mechanics: on the question of action at a distance, for example. We cannot apply double standards here: if the anomalies in Aristotelian mechanics on the issue of the resultants of constant forces make classical mechanics a better theory of this issue, then the anomalies in Newtonian mechanics on the issue of action at a distance make Aristotelian mechanics a better theory of *this* issue. But assessments of this kind would be trivial. Some of our best theories give rise to anomalies in areas in which the theories that they have superceded had no problems. It is always possible to overcome the isolated anomaly. Indeed, the best way to overcome them is to find out why they arise in the first place, and this involves reference not to other discourses but solely to the discourse in which they arise, and it is because of this that a comparative assessment of the individual explanations is precluded.

This does not, however, preclude any kind of assessment. Indeed, two kinds of assessment are possible. First, we can assess the explanatory structures of the discourses in which particular explanations are produced. Secondly, if we can establish that there is an overlap of reference between the domains of investigation of the discourses concerned we can then make a comparative assessment of the explanatory structures of those discourses. It may be worthwhile at this stage to indicate briefly what is involved in these kinds of assessment.

I propose to examine two explanatory structures in Part II. For the sake of simplicity, these can be called the explanatory structures of 'Aristotelian' and 'Galilean' mechanics. These are two different discourses because they have different explanatory structures. The two explanatory structures will be analysed independently. On the basis of this analysis we shall ask what counts as an explanation in these discourses, whether explanations of the required kind can be given in principle, whether they can be given in fact, and if so under what conditions. This should enable us to identify different kinds of explanatory failure. For example, a discourse which cannot generate, in principle, any explanations of the kind required in that discourse is subject to an explanatory failure of a different and more serious kind than one in which explanations of the required kind can be given in principle but not in fact. In giving an account of explanatory failure in this fashion we shall be assessing the explanatory structures of discourses. An account of this kind is an 'internal' one: no reference need be made to other explanatory structures (and still less to what we currently think the best explanations of the phenomena in question are). This means that whilst it is true that explanatory structures are constrained, in an important way, by what the domains of investigation of discourses are, the analyses of the explanatory structures of any two discourses are independent of the way in which the domains of investigation of those discourses are related. This means that comparative problems of reference simply do not arise. In this sense, it makes no difference whether we are discussing the explanatory structures of Freudian psycho-analytic theory and Newtonian cosmology, or Aristotelian cosmology and Newtonian cosmology: our account of each of these would be independent. In this case, meaning-variance and reference-variance would not arise as problems because we would not be concerned with the relations between domains of investigation.

Now we can (at least in theory) compare any two discourses at the level of the kinds of explanatory failures which they incur, but such a comparison would be pointless and uninformative unless we could construe the discourses concerned as providing competing accounts. It is *here* that the problems of the relations between domains of investigation comes in, for in determining whether this latter situation holds we would have to establish whether there was any overlap of reference in the domains of investigation concerned.

To sum up, in the case of discourses which are differentiated solely in terms of their explanatory structures, and where two variables — what is being explained and what counts as an explanation — are operative, there are three possible modes of assessment, depending on the circumstances which hold. First, where what is being explained is the same, and where the explanations are produced in the same discourse, one can compare individual explanations. Secondly, where what is being explained is different and where the explanations are produced in different discourses, we can only (informatively) assess the explanatory structures of the discourses individually. Thirdly, where what is being explained is constant and where it is being explained in different discourses we can assess the explanatory structures of these discourses comparatively: the degree to which this is possible depending on the extent of the referential overlap in the domains of investigation.

Finally, I should point out that although I have concentrated on assessment in this section, assessment is not the only, or even the most important, motive for analysing explanatory structures. It is crucial that we be able to give a proper account of explanatory failure, not just for the purposes of assessment but also so that we might be able to recognise the kinds of problems that would have to be dealt with in a discourse if particular kinds of explanatory failure are to be overcome. It is for this reason that a theory of explanatory structures is being proposed, and we must now consider what is involved in this theory in detail.

Notes: Chapter 1

1 Kuhn's account was originally presented in his *Structure of Scientific Revolutions*. 'Revised' versions of the theory are to be found in his 'Reflections on my Critics' and 'Second Thoughts on Paradigms'. In 'Second Thoughts', paradigms are replaced by 'disciplinary matrices', but none of the general problems which I shall raise are overcome by this substitution.
2 Cf. Kuhn, *Structure of Scientific Revolutions*, pp.120 and 128 (epistemology), p.85 (psychology), p.56 (optics).
3 *Ibid*, p.94.
4 Cf. Kuhn, 'Reflections', p.232.
5 Kuhn, *Structure of Scientific Revolutions*, p.150.
6 *Ibid*, p.181.
7 Cf. Lakatos, 'History of Science and its Rational Reconstructions', 'Falsification and the Methodology of Scientific Research Programmes'; Lakatos & Zahar, 'Why did Copernicus supersede Ptolemy?'; Zahar, 'Why did Einstein's Programme supersede Lorentz's?'.
8 Lakatos, 'Falsification', p.133.
9 *Ibid*, p.132.
10 *Ibid*, p.132.
11 On some versions of a radical meaning-variance theory — such as Feyerabend's — domains of investigation may well serve to differentiate research programmes, since Feyerabend seems to consider that variance of sense brings with it variance of reference (and this is something that we shall question later). That Lakatos does not accept such an account is clear from his remark that 'we can make [incommensurable theories], by a dictionary, inconsistent and their content comparable' (*ibid*, p.179). I point this out simply to show that, on Lakatos' account of incommensurability, research programmes cannot be differentiated in terms of their domains of investigation. I am not claiming that this account of incommensurability is right. In fact, I find it very obscure as it stands. The scope of the statement is left unspecified, and it surely cannot refer to *any* 'incommensurable theories' — such as psychoanalytic

theory and Newtonian cosmology. We therefore need some further specification by which to decide which 'incommensurable theories' can be 'made' comparable and which cannot be 'made' comparable, and I can find no such specification in Lakatos' work.

12 *Ibid*, p.133. The original has 'hypothes*is*', which is surely a misprint.

13 Cf. Putnam, *Philosophical Papers*, I, pp.174-197.

14 Lakatos, Falsification', p.116.

15 Feyerabend, 'Zahar on Einstein', p.27.

16 Zahar, *op cit*.

17 Lakatos and Zahar, *op cit*.

18 For a fuller discussion of this issue see Gaukroger, 'Bachelard and the Problem of Epistemological Analysis'.

19 Hesse, *Structure of Scientific Inference*, p.64. Hesse has managed to refine the Fregean account of sense and reference considerably. There can be no doubt that the idea of 'sense', in particular, does require refinement. When I speak of *a* Fregean account, this is to be distinguished from Frege's own account.

20 Putnam, *op cit*, II, Essays 6, 9, 11, 13, and especially Essay 12.

21 *Ibid*, pp.235-6.

22 Since the problem of essentialism will be dealt with in detail in Part II, where its role in early physics will be considered, I shall not discuss it further here. It is perhaps worth noting, however, that there is considerable disagreement throughout the development of physics on what exactly essential accounts are supposed to look like. In particular, I think that on Aristotle's account of essences, contemporary atomic explanations of gold would be unacceptable on the same grounds as Greek atomist accounts were considered unacceptable: to explain the properties of gold in terms of the properties of atoms is only to push the problem one step back since, for Aristotle, to explain the properties of matter we require something different in kind from matter (an immaterial 'principle'). It is also noteworthy that essentialists have traditionally found it necessary to introduce some special 'intuitive faculty' for grasping essences (or, alternatively, to posit some special 'mental state' that we are in when we grasp essences). Something of this kind is clearly required but Putnam (rightly) argues against the idea of special intuitions, thereby leaving himself with no basis on which to distinguish between essential and non-essential principles of laws.

23 Ellis, 'Newton's Laws of Motion', p.49.

24 Cummins, 'States, Causes and the Law of Inertia', p.29. Cummins does not claim that this principle holds universally, and he suggests (p.33) that it may break down in quantum mechanics.
25 *Ibid*, pp.28-9.

PART I

CHAPTER 2

ONTOLOGY AND REALITY

§1 Ontological Discontinuity

IN section 2 of the last chapter I said that all explanatory structures have an ontology and a domain of evidence, and that all theoretical discourses have an explanatory structure. The primary purpose of this chapter is to clarify the ideas of an *ontology* and a *domain of evidence*. In the course of our discussion we shall meet the cases of discourses which do not operate with a domain of evidence, and hence do not have an explanatory structure, and we shall call these 'atheoretical'. I hope to be able to resolve some of the problems surrounding ontologies and evidence in this chapter so that the way will then be open to an examination of the link between ontologies and domains of evidence in chapter 3.

The ontology of a theoretical discourse is that primary structured set of kinds of entity in terms of which explanations can be given in that discourse. A necessary condition for something's being an *entity* is that it would figure as a bound variable in a complete formalisation of that discourse.[1] Entities are of the same kind if the same set of predicates is applicable to them (or, more strictly speaking, to the terms designating them); they are of different kinds if the same set of predicates is not applicable to them. A predicate is applicable to a member of a kind if and only if either its affirmation or denial is true of that member, except where it is true *simply in virtue of the entity's being a member of the kind*: in which case it is inapplicable. (It may be true, for example, that this particular table is not green, but this is not due to its being a member of the class of tables; it is true that my concept of a table is not green, but this *is* due to its being a member of the class of concepts. Hence the predicate

39

'green/not green' is applicable — on the definition proposed
— to tables but not to concepts.)

Insofar as any predicate can be used to classify kinds, this
classification is conventional. But insofar as only particular
predicates will be used, within any particular discourse, to
classify kinds, the classification is not an arbitrary one.
Similar considerations apply to the individuation of entities
within kinds. There is no *a priori* reason why there should be
one set of (non-logical) necessary and sufficient criteria by
which entities of any particular kind can be individuated.
Observable, 'middle-sized', spatio-temporal entities have, as
part of their criterion of individuation, the condition that
they can be reidentified through time.[2] But this is no part of
the criterion of individuation of electrons, for example:
electrons can be localised but they cannot be reidentified.

Provisionally, we can say that a set of kinds of entity
which is primary is a set whose members — kinds of entity —
are irreducible to one another, and irreducible to other kinds
of entity: they are independent kinds. On a similarly provisional basis, we can also say that for a set of kinds of entity
to be the ontology of a discourse, all explanations which can
be given in that discourse can be given in terms of this
primary set, or in terms of kinds of entities reducible (in that
discourse) to this primary set or members of it.

One of the main aims of this chapter is to examine what is
involved in the claim that discourses can be ontologically discontinuous with one another. In particular, we shall be concerned with the status of those discourses whose ontologies
are distinct from that of sense experience. The ontology of
sense experience will be taken, primarily, to be that set of
entities which are observable in principle, that is (Euclidean)
three-dimensional bodies with a continuous endurance in a
linear and asymmetric time.[3] The thesis that discourses can
be ontologically discontinuous with sense experience
amounts to saying that the study of particular phenomena
may require the postulation of the existence of a set of
entities to which we have no access in sense experience, and
where this set of entities cannot be reduced to those which
we do have access to in sense experience.

In considering these issues, one of the main questions

which arises concerns the relation between particular ontologies and 'reality'. The problems involved here can best be
illustrated by considering a question posed by Eddington.[4]
He asks whether the *real* table is the extended, permanent,
coloured and 'substantial' thing presented in perception, or
whether it is the 'scientist's' table which is, amongst other
things, colourless and well over 99% vacuum. His question
can be re-phrased as follows: should we ascribe reality to the
kinds of entity with which we are familiar through observation, or should we ascribe reality to the kinds of entity in
terms of which physics operates? Assuming that reality can
be ascribed to at least one of these ontologies, and taking the
'or' here as inclusive (that is, *vel* rather than *aut*), there are   *The 3 options*
three possible replies to the question. They are: (1) only
observable entities are real; (2) only the entities of physics are
real; and (3) the entities of both sets are real — or, at least,
they are both candidates for this status. On the first two
options there is one basic ontology, which is that of 'reality',
to which all other ontologies are in some sense secondary.
They differ in their accounts of what this basic ontology is.
The third option can be construed in several ways, but
whichever way it is interpreted what is being denied is that
the entities of sense experience and those of physics *must
compete* for reality.
   We shall examine these options in detail, since it is clear
that, depending on how we construe 'reality', some discourses will not explain 'reality', or they will offer explanations in terms of entities which are not 'real'. It is therefore
of some importance that the question of the relation between
ontologies and reality, and the question of the criteria by
which something is established as being 'real' in a discourse,
are dealt with. The conclusions that we come to on these
issues have an important bearing on how we treat the
problem of "idealisations", which is something we shall be
concerned with when we come to consider Galilean
mechanics, later in the book.

§2   Observability

   In essence, the first option can be stated as follows. The
ontology of 'perception' constitutes 'reality', whereas the

(unobservable) constructs of physical theory, for example, are in some sense secondary. On this account, the ultimate objects with which physics deals are pieces of homogeneous material whose quantity as an aggregate remains constant but whose spatial distribution changes in a continuous fashion in accord with laws which it is the business of mechanics to formulate.

One rather weak version of this option is the thesis that only those things which are in fact observable are real. G. Maxwell[5] has successfully criticised this thesis on the basis that things that we cannot in fact observe now may be observable later with the development of more sophisticated instruments. A more interesting version of the first option is that according to which those entities are *real* which, although they may be too small to be perceived, nevertheless share their fundamental properties in common with "middle-sized" objects. By fundamental properties, I mean such things as weight, extension, determinate position, duration in time, and so on.

An account of this kind was commonly (and unsuccessfully) put forward during the seventeenth, eighteenth, and nineteenth centuries as an epistemological and ontological basis for classical mechanics. Its main features can be summarised as follows. Matter, space and time[6] are the three fundamental ontological kinds: any other kinds of entity in physics are ultimately reducible to these. Matter is defined simply as full space, and on this definition the primary qualities of matter — other than its fullness — are those which it shares with the space it occupies. Space on this account is three-dimensional and Euclidean, and it is conceived as a homogeneous container which is independent of its physical content, physically inert, infinite in extent and infinitely divisible. Since matter is full space, and since in saying it is full we cannot admit of degrees of fullness, it follows that matter is impenetrable and inelastic. If matter were penetrable the space it occupies would have to become 'more full' on penetration. Similarly, if matter were elastic the space it occupies would have to become more or less full whenever this elasticity was exhibited. This follows from the isotropy and homogeneity of Euclidean space. Finally, since the fact

that motion occurs cannot be reconciled with the idea of a
(homogeneous) plenum — despite Descartes' ingenious arguments to the contrary — it is clear that the existence of empty
space must be posited.

What we have, then, is matter, which is 'substantial' and
has weight, moving in empty space which is 'insubstantial'
and weightless. Many of the paradoxes to which the first
option seems to be committed disappear on this account. The
substantiality of the table, for example, is due to an aggregate of ultimate pieces of matter. These pieces of matter are,
by everyday standards, 'super-substantial' but there is no
longer anything counter-intuitive in the fact that a middle-
sized body is both substantial *and* 99.9% vacuum. The table,
and the matter from which it is constituted, have their fundamental properties in common. The table itself is just a
particular concentration of matter in a particular place, the
degree to which it is substantial depending on the degree of
concentration. The table is coloured, of course, whereas the
ultimate pieces of matter are not. But this is not a problem
since colour can be treated simply as a 'secondary quality', as
a psychic addition of the perceiving mind. This does not
render the ultimate pieces of matter imperceptible in
principle since they are still extended, or even in fact since
they still have a weight. Colour is not as important here as
some philosophers have thought. The most hard-line posi-
tivist would not deny the existence of colourless gases just
because we cannot see them. *A fortiori*, there is even less of a
problem with colourless pieces of matter which, unlike
colourless gases, have a fixed extension.

The interest of an account such as this derives from the
fact that it is not dependent on considerations of size. Those
'theoretical constructs' of physics which, while they may be
localisable, have neither determinate extension — if they
have any extension at all — nor determinate position, are not
denied reality because they are too small to be seen, but
because the kind of properties ascribed to them render them
unobservable in principle. Nevertheless, entities which are
unobservable in principle still pose something of a problem,
since they seem to play an important and indeed indis-
pensible role in physics — or, at least, in physics after the

development of field theory and the field interpretation of mass-points. The main way in which this problem has been dealt with by proponents of the first option is by the adoption of an instrumentalist conception of the function of theories. In instrumentalism, theories are conceived as nothing more than logical instruments for ordering and predicting experience. Those entities which are unobservable in principle ('ideal objects' etc) and to which the theories must refer in order that they might perform this task in the most simple and comprehensive manner, do not really exist. Since theories function primarily as rules on the instrumentalist account, they cannot be characterised as being either true or false.

If a fully 'instrumentalist' position of the Machian kind[7] — according to which theories are simply a 'convenient shorthand' for a class of observation statements — is taken up, then certain debates, which are actually concerned to determine whether particular entities or kinds of entity are merely conventional or not, are rendered unintelligible. Consider, for example, the debate over the status of potential energy in classical mechanics. On the instrumentalist thesis, the question of whether potential energy exists in its own right, like kinetic energy, or whether it is merely a mathematical convention for balancing equations, and is really only disguised kinetic energy, can only be a bogus question. A similar example is that of the debate over fields of force and gravitational functions as these figure in the sets of equations in mechanics from Laplace and Poisson onwards.

Secondly, examinations of the role of instrumentalism in particular debates have often indicated its retrogressive role. Feyerabend, for example, has argued that the instrumentalist arguments of Proclus and some early seventeenth century astronomers, on the one hand, and Bohr, on the other, can be shown to have blocked the development of the heliocentric and quantum theories respectively.[8] Similarly, it is difficult to see how work on molecular theory could have come to such a successful experimental conclusion had the late nineteenth and early twentieth century debates on molecules — debates which were concerned as much with the meaningfulness of asserting the existence of molecules as with the nature of molecules — been decided in favour of the instrumentalists. The resolution of these issues did not merely involve experimental 'observation' or the production of new correlations. Nye, in a detailed study of Perrin's work, has noted that the instrumentalist theses of Mach and Duhem had to be dealt with before the experimental project in which Perrin wished to engage could be conceived properly. She argues that a study of Perrin's work reveals 'that the entire question of molecular reality involved the problem of procedure and a redefinition of the aims of scientific endeavour'.[9]

One motive for equating observability and reality has traditionally been the assumption that there is a distinction of kind to be made between theoretical and observation terms. On this account, there are terms at two levels in a theory: the observation terms are taken to be unproblematic and the theoretical terms are construed as requiring explanation by reference to observation terms. The relation between the two levels is then formulated by means of a set of translation procedures.[10] It is clear, however, that observation is simply not possible without some kind of conceptualisation. When we observe something we observe it as being something of a certain kind. This identification is dependent on some conceptualisation, some classification. No demonstrative reference could be successful unless there were some shared principles of classification. For example, if I am asked to count the number of items in a certain area, I must restrict myself to a single system of classification — if I am counting the number of things in a room I cannot include a chair, wood, legs, molecules and oblong shapes in the same total. Thus observation is necessarily a conceptual activity insofar as it involves a system of classification and criteria by which things of a certain class are identified as being such, and thereby differentiated from other things of that class and from things of other classes.

Even apart from these issues, the notion of observability is very problematic. In visual perception, for example, those things which emit electromagnetic waves of a certain wavelength range are observable, and those things which emit radiation with wavelengths outside this range are unobservable. This is a purely contingent feature of our perceptual apparatuses. It is quite legitimate to distinguish those entities which are (visually) observable in principle from other kinds of entity, but to confer on the first kind of entity an ontological priority,[11] to say that only entities of this kind are *real*, is wholly unwarranted. It is a blatant *non sequitur* to argue from purely contingent features of our visual apparatuses to what exists.

§3   Evidence

The thesis of the second option is that the entities of physical theory are the only real entities, and that the entities of perception are not real. Sellars gives a clear statement of this position when he claims that 'according to the view I am proposing, correspondence rules appear in the material mode as statements to the effect that the objects of the observational framework *do not really exist — there really are no such things*'.[12] A similar thesis is put forward by Feyerabend,[13] who argues that science replaces 'common sense' *and* that common sense is rendered dispensible in the process. This argument that the 'primitive' ontology of sense perception has been replaced by (what Sellars calls) 'more sophisticated ontologies' with the growth of physical theory can take two forms. Sellars[14] has argued that the common sense ontology has a certain pragmatic validity and may be retained for practical purposes; Feyerabend, on the other hand, argues for total replacement.

Feyerabend's thesis — if not Sellars' — rests on the construal of common sense as a theory with its own 'observation language'. That is to say, the analysis which is put forward to deal with physical theory is also designed to cover common sense without distinction. In Feyerabend's work this is linked with two particularly important theories about observation. These are: (1) We cannot choose between alternative theories on the basis of observation since observation is itself theory-dependent;[15] (2) If a theory is inconsistent with an observation it should not be reviewed or discarded simply because of this; on the contrary, the observation should be reinterpreted to fit the theory.[16]

The first thesis rests on the idea that what counts as a veridical observation, what is observed, and how it is observed, all depend on the theory in which the observation is made. This means that there need be no common parameter of observation between two 'conflicting' theories. Feyerabend attempts to show in detail how there cannot be an observation which is independent of a theory by considering Galileo's treatment of free fall. For purported 'neutral observations' he substitutes 'natural interpretations':

> 'One can distinguish between sensations and those "mental operations which follow so closely upon the senses", and which are so firmly connected with their reactions that a separation is difficult to achieve. Considering the origin and effect of such operations, I shall call them *natural interpretations.* '[17]

In his discussion of Galileo's law of falling bodies, Feyerabend notes that the classical argument from falling stones seems to refute the Copernican theory of the earth's motion. The argument from falling stones rests on the 'observation' that stones fall with a perpendicular, rectilinear motion; that is to say, if we drop a stone from the top of a tower it will fall straight to the base of the tower. Now if the earth were moving, so the argument goes, this would not happen, since during the time between the stone's being let go and its reaching the ground the earth would have moved, and hence we would expect the stone to reach the ground at some distance from the base of the tower. This part of the argument is based on the 'observation' that if, for example, we were sitting on a bridge — which simply counts as some suitable stationary object — and if we drop a stone just as the bow of a ship passes under the bridge, the stone will not land on the bow but at a place some distance behind the bow, for during the time in which the stone takes to fall, the ship will have moved a certain distance. Feyerabend argues that Galileo replies to this objection by presupposing the theory that it is supposed to refute (the Copernican theory), turning the objection round, and then using it to discover the natural interpretations that exclude the motion of the earth. That is to say, Galileo first asserts the motion of the earth, and then enquires what changes will remove the apparent contradiction. The form this argument takes in Galileo is well known: he argues for the relativity of all motion by pointing out that notwithstanding the bridge example, a stone dropped from the top of a *mast* of a moving ship will fall to the bottom of the mast. Hence, what Galileo actually does, Feyerabend argues, is to discover the particular natural interpretations involved by assuming the thesis objected to, and then to undermine these natural interpretations by using the senses as instruments of exploration only with respect to the reality of *relative* motion. In doing this, Galileo introduces a new observation language, by replacing the conceptual system in which motion is an absolute process which can always be perceived — since it always has effects, effects on our senses included — by a new conceptual system in which all motion is relative. In the new conceptual system, not all motions can be perceived. Thus, Galileo can argue that the free fall of a body really describes a curved motion, but that we can only see the rectilinear component of this because the rectilinear motion is the only one in which we do not share. This, in turn, enables him to argue that the motion of a falling body is rectilinear only with respect to an observer who is stationary relative to the fall.

Feyerabend's conclusion is that if a theory is inconsistent with an observation then the observation should be reinterpreted to fit the theory. This, he claims, is what has occurred when major breakthroughs have been made in physics. What is problematic in this account, however, is the treatment of common sense (and sense perception) as a theory, and the corresponding claim that, as such, it is replaceable by a different theory.

To say that this is problematic is not to deny that there exist theories, or even co-ordinated bodies of theory, which can be called 'common sensical'. Nor can it be denied that these theories constitute schemas which determine what is to count as a veridical observation in any particular circumstances. The theory that the earth is stationary about its own axis is 'common sensical', for example. The observations which we count as veridical in everyday experience both suggest and confirm such a theory: we do not feel the forces usually associated with motion, and birds can hover in the air above us without their being a muscular effort required on their part in order that they might keep in the same position with respect to us. Further, this common sense theory has been replaced by a quite different theory, and the observations mentioned have been wholly 'reinterpreted' to support a theory which, *prima facie*, they tend to disconfirm.

It is clear, however, that if 'common sense' is to be construed as theoretical at all then it cannot simply be regarded as one large theory. Rather, it must consist of a number of quite different kinds of theory, some informed by developments in physics (and other areas) and some not. Further, if it does consist of such theories these are rarely explicit. The idea of non-explicit theories raise a more fundamental problem concerning the extent to which 'common sense' can be treated as homogeneous. The problem is that there are some elements in 'common sense' which can be called 'theoretical' only in a very strained sense. In order for something to count as a *theory*, it does seem that we must at least know what would count as evidence for and against it. In the case of the earth's diurnal motion, for example, this condition is satisfied.

There are cases, however, where it becomes very difficult to regard 'common sense' as theoretical, even non-explicitly theoretical. Consider the existence of tables, for example. Does and could the acceptance of molecular theory change conceptions about the existence of tables? With regard to the first part of the question, this acceptance does not change behaviour towards tables. Those who claim that the ontology of 'middle-sized' objects has been replaced by a 'more sophisticated' ontology do not treat tables as if they did not exist — by attempting to walk through them for example. Someone who did this could safely be regarded as insane. But someone who did not do this could not reasonably claim 'pragmatic' motives. If there are reasons for not attempting to walk through tables — which are, after all, 99.9% vacuum — these are not pragmatic in any usual sense of the word. Bearing this in mind, it does seem that molecular theory is not an *alternative* to 'common sense' ideas about tables in the way that the theory of the earth's diurnal motion is an alternative to 'common sense' ideas of the earth being stationary about its axis. An indication of the difference is that in the latter case we know what would count as evidence against the common sense idea, whereas in the former case it is unclear what would count as evidence against the existence of objects such as tables. Indeed, it is difficult to know what would count as evidence *for* the existence of objects of this kind. We can have evidence *that* something is a table, but what kind of evidence can we have for tables?

Consider the following argument of Austin's:

> 'It is not the case . . . that whenever a "material object" statement is made, the speaker must have or could produce evidence for it. This may sound plausible enough; but it involves a misuse of the notion of "evidence". The situation in which I would properly be said to have *evidence* for the statement that some animal is a pig is that, for example, in which the beast itself is not actually on view, but I can see plenty of pig-like marks on the ground outside its retreat. If I find a few buckets of pig-food, that's a bit more evidence, and noises and the smell may provide better evidence still. But if the animal then emerges and stands there plainly in view, there is no longer any question of collecting evidence; its coming into view doesn't provide me with more *evidence* that it's a pig, I can now just *see* it is, the question is settled.'[18]

The point being made here has an important bearing on our present discussion. I have argued that it is a necessary condition for something's being counted a theory that we know the kind of thing that could count as evidence for or against it. Austin's example raises a problem in that the case discussed does not seem to be one which fulfils this condition. That is, at least as far as Austin's example is concerned, existential statements about pigs are *atheoretical*. This does *not* mean that in order to identify a pig we do not need to apply concepts. In picking out something as being of a certain kind, concepts are always required. What is at issue is whether all procedures which involve concepts can be called 'theoretical'. If we wish to restrict the use of the term 'theoretical' to statements which can (in principle) be assessed in terms of evidence, then it does seem that we must distinguish these statements from statements of the kind where it would be inappropriate to ask for evidence. Nevertheless, this distinction, however clear cut it may be conceptually, does not have such a clear cut application. The problem is that we can always think of situations where evidence would be required for any existential statement, and this undermines the idea that there are absolutely atheoretical statements. It does not undermine the idea that statements may be atheoretical with respect to a particular context, however, and we can obtain some clarification of the issues involved by examining the role which 'context' plays here.

For any given statement, we can ask for the kinds of situation in which it would be appropriate to ask for evidence for the claim made by that statement, and we can ask for the kind of evidence that would then be called for. In the case of existential statements in fundamental particle theory, for example, questions of evidence are always appropriate. In the case of existential statements about tables, the situation is rather more complicated. I do not *infer* the existence of a table from seeing a table. Under normal circumstances we do not refer to evidence to establish the existence of a table. In atomic physics, it is essential that we make such a reference. The distinction between these two cases hinges on the expression 'under normal circumstances'. Before we attempt to elucidate this qualification, it may be worthwhile to examine briefly a more problematic kind of case which also appears to have a claim to being 'atheoretical'.

The problems of evidence which arise in the case of existential statements about 'middle-sized' objects are not unique. Other cases are even less clear cut. The statement 'the sun revolves around the earth' is usually considered to be a theoretical statement. This statement has been replaced by a different theoretical statement: 'the earth revolves around the sun'. Now the statement 'the sun rises at dawn' appears, on this basis, to be theoretical also, in that it is a consequence of the geocentric theory. Moreover, it appears incompatible with the heliocentric theory. But there is a sense in which my saying 'the sun rises at dawn' is atheoretical, and in this sense such a statement does not commit one to the heliocentric theory. It is instructive here that such statements are still in common use long after the acceptance of the earth's motion around the sun. They have not been replaced by statements more in keeping with the heliocentric theory because they are *not in competition* with this theory.

This does not mean that they are not replaceable, only that — because of the role they have — they are not replaceable by theoretical statements. This is indeed the crux of the matter, and at this point we can begin to appreciate the problem of existential statements about things such as tables. Statements of this kind depend on such things as the physiological structure of our sensory apparatuses, the structure of language, our position in the universe, and so on. If these were different — if, for example, we perceived radio waves and not light waves — then we might expect these statements to be different. We might also expect severe problems to arise with statements of this kind if we lived in another part of the universe. In the presence of massive gravitational fields, the behaviour of tables may be such that questions of evidence may continually arise. Questions of evidence also seem to depend to some extent on culture (although much less radically so than in the other cases that we have mentioned). Now it is changes of this kind that would call for a replacement. The situation with regard to theoretical discourses is quite different. Changes in our physiology, or changes in our position in the universe, do not change the adequacy of explanation in physics, although they may change their usefulness for any particular case.

The fact that different kinds of replacement are involved, depending on the kind of statements (and ultimately on the kind of discourse) one is considering, has been noted by Sellars:

> 'Not all subject-matter dependent universal propositions to which common sense is committed are properly characterised as *beliefs*. There are many principles about physical objects and the perception of them ("categorial principles" they may be called) which are constitutive of the very concepts in terms of which we experience the world. And while I agree with Feyerabend that this does not exempt them from criticisms and possible replacement, it is, as I see, incorrect to compare this possible replacement to the replacement of a corpuscular by a wave theory of light, or dephlogistication by oxidation.'[19]

Sellars does not present any clear cut solution to the problem of distinguishing between the different kinds of replacement.

This is a complex issue[20] and one that we shall not go into here, since our prime concern is with theoretical discourses — which always operate with a domain of evidence. There are, however, four points worth bearing in mind. First, statements about which it is inappropriate to ask for evidence have no explanatory value — this is not a failing of such statements, it is simply that they serve a particular kind of function which is not that of explaining. Secondly, the claims of such statements are not purely 'observational' — they depend on concepts like any other statement, but this does not thereby render them theoretical (in the sense defined). Thirdly, there are no atheoretical statements *simpliciter* — particular statements are atheoretical only with respect to a particular context, and when this context changes the nature of the statements may change. Fourthly, one of the most important ways in which atheoretical statements are to be distinguished is by the conditions under which they are replaceable.

§4   The Concept of an Ontology

Up to this point, I have tried to establish that ontologies need not compete for reality. Whether particular ontologies do compete for reality is a question which can only be decided by examining these ontologies and the discourses in which they are put forward. Here we are concerned simply with what the relations between ontologies *can* be. That is to say, we are concerned purely with possibilities; what the relations between particular ontologies *are* is something which requires a different kind of treatment which we are not yet in a position to give.

Although we have been concerned with problems of ontology in this chapter, we have not yet discussed the question of what an ontology is in any detail. Before we can do this there is a problem which we must deal with. It has been claimed by Nagel that theories — or theoretical discourses — not only determine the criteria for something's being real, but that they also determine the criteria by which we decide what it means to say that something is real.

If we say that atoms, tables and galaxies exist in different ways — meaning by this that the word 'exists' here is being used in different senses — then the problem of what exists becomes a purely verbal matter. Nagel, who puts forward an account of this kind, construes the realism/instrumentalism debate in terms of a purely verbal dispute.[21] There are two reasons why this will not work. First, to propose different senses of existence only pushes the problem one step back. If the different meanings are wholly incommensurable then we are not talking about existence at all, and Nagel is simply not dealing with the issue in question. If they are not wholly incommensurable then we might reasonably expect to be told what they all share in order that they might all be called 'existence'. It is at the point of overlap that the original problem returns.

Now it might be argued against this objection that in saying that 'existence' has different senses, what is being claimed is that 'existence' is a family-concept. That is to say, the different senses of existence are not wholly incommensurable but this does not mean that they all have something in common. This reply is plausible, but there is a second problem which it does not overcome. It is important that we be able to distinguish (at least in principle) fictional and 'conventional' entities from those which are not of this kind. To my knowledge, the only account which attempts both to distinguish kinds of existence and to give an account of the difference between fictional and non-fictional entities is that of Meinong, who discriminates between *existence* and *subsistence*. The logical difficulties and paradoxes inherent in this account are notorious.[22]

In arguing against the idea that there are legitimately different senses of the word 'exists' — or, at least, in denying that such an idea is of any use in explaining what is at issue in the realism/instrumentalism debate — there are two kinds of case that I do not want to rule out. Both these cases could be construed in terms of things existing in different senses but such a construal would be misleading; moreover, they are clearly different from the kind of case that we have just considered. The first case is that in which we accept a complete reduction of tables to molecules (or fields or what-

[handwritten margin note: Why assume that it is in virtue of shared or common features that the term is applied to all of these? Aristotle's categories.]

ever) and where we do not thereby wish to deny the existence of tables. In this case, to claim that molecules and tables exist in different senses would simply be a misleading way of saying that molecules have an independent existence with respect to tables, but not vice versa: that is, molecules can exist without the existence of tables but tables could not exist without the existence of molecules. This can be considered as a case of strong ontological priority.[23] The second case is that where weak ontological priority is involved. We might include matter and space in our ontology, for example, but consider that whereas there can be empty space, there could be no matter without the prior existence of space. We shall come back to this case shortly; for the moment it is sufficient to note that the claim that matter and space exist in different senses would be unwarranted. To say that matter can exist only if space exists need not render the meaning of 'existence' equivocal.

I now want to turn to the crucial issue: the problem of what exactly it is that we establish the reality of. I shall argue that it is the existence of entities of particular kinds within particular ontologies which can be established, where 'within particular ontologies' is the operative phrase.

This qualification is important, since it distinguishes the position I want to defend from that taken up by Mellor, for example, when he claims that 'the Universe, so far as we have reason to believe, contains as independent entities both galaxies and electrons — *and* tables and men'.[24] What is problematic about this claim is the presupposition that entities can simply be lifted out of their ontologies and listed. To posit the independent existence of electrons and galaxies is not simply to posit the existence of two different kinds of entities, but two different kinds of ontologies which may conflict in that one may also include space and time as independent entities, for example, whereas the other may also include space-time as an independent entity. *Prima facie* compatibility between selected entities from two ontologies does not mean that the ontologies are compatible. The issues here are complex, and they can best be approached by considering what sense it makes to speak of an ontology being 'complete'.

Ontologies usually contain more than one kind of entity. The ontology of Classical Atomism — or, as Čapek characterises it, the 'corpuscular-kinetic' schema[25] — for example, includes matter, time and space. A different ontology would be that which includes fields, space and time, or simply space-time. However, and this is the crucial point, ontologies do not just contain kinds of entity, they contain kinds of entity in specific relations to one another. In Classical Atomism, for example, matter is 'in' space and time, but space and time are not 'in' matter. The way in which the kinds of entity in an ontology are related to one another is at least as important as the kinds of entity which are included in that ontology. It is the kinds of entity, together with the relation between these kinds of entity, which serve to *unify* an ontology. For example, if a Classical Atomist mechanics is put forward to account for motion, and if it includes only matter and space as independent entities in its ontology, then its project is unrealisable in principle (on internal grounds). If we wish to keep matter in the ontology, then one of (at least) two things may be done. First, time could be introduced as an independent entity: that is, as something which exists independently of matter and space. On this account, the original ontology would be construed as being incomplete. An alternative would be to remove space from the ontology and to treat it as a dependent entity: this would be to conceive of the existence of space as being dependent on the existence of matter. On this account, space and time would be construed purely relationally, and the original ontology would be conceived as being based on a mischaracterisation of dependent entities.

It will be clear even from these two very simplified cases that the relations between kinds of entity in an ontology is complex. Dependence, in particular, is neither a simple nor even a univocal relation. In the Classical Atomist ontology for example, we can say that the three kinds of entity are matter, space and time. These each have an independent status in that none is reducible to the others. Nevertheless, while space and time are conceived of as requiring neither each other nor matter for their existence, matter cannot exist without the (logically) prior existence of space and time on this

Cp. Hempel on the 2 kinds of internal princi- ples of a theory account.[26] Furthermore, the situation is complicated by the fact that these peculiar dependencies are in part determined by other features of the discourses of which they are part: we shall meet some examples of this when we come to consider the case studies. For the moment, it is sufficient to note that the question of the *unity* or *coherence* of an ontology is not something which can be decided by a simple inspection of that ontology, even if this inspection is carried out on the basis of Quinean criteria for ontological commitment. This would not be viable even in the simplified cases that we have considered.

I am arguing that ontologies do not simply consist of kinds of entity — they consist of kinds of entity bearing certain complex relations to one another. It is these two features, together, which go to determine the coherence of an ontology, and it is this coherence that makes it meaningful to ask whether a particular ontology is incomplete, or whether it contains kinds of entity which are wrongly (that is, in this case, 'inconsistently') construed as being irreducible to other kinds of entity in the ontology. These questions may be difficult to answer in any particular case, but this does not render them any the less meaningful or important. Further, it is this feature of ontologies — the fact that they consist of sets of entities *in complex relations to one another* — that renders the selective listing of entities from different ontologies such a bizarre enterprise. If the listing is simply produced as a provisional one, in the expectation of some future reduction, then it becomes more plausible, since its peculiarity would then be understandable, in that the list could be taken as picking out those kinds of entity which currently cause the greatest difficulty for any reduction project.

Notes: Chapter 2

1   If, with Quine, we are only prepared to allow an extensional predicate calculus, then ontological commitment must be restricted to the bound variables of a first-order quantification (cf. Quine, *Set Theory*, esp chs. 1 and 2, and *Word and Object*, chs., 6 and 7). If, with most other logicians, we allow

intensional calculi, then ontological commitment in the case of bound variables of higher-order quantification cannot be ruled out. There are two points to note here. First, ontological commitment cannot be restricted to bound variables under the scope of an existential quantifier, as Quine sometimes seems to suggest. As Dummett (*Frege*, p.476) points out, '$\forall x\, \exists y\, A\, (x,y)$' must have as much existential import as '$\exists x\, B\, (x)$'. Secondly, whatever level of quantification we are interested in, we must allow, or at least allow the possibility of, many-sorted quantifications. Higher-order predicate calculi are of course many-sorted systems, but they are many-sorted systems of a special kind in which the membership relation has a unique status. While all higher-order calculi are many-sorted systems, not all many-sorted systems are higher-order calculi. Elementary geometry, for example, which deals with points, lines and planes, is a many-sorted theory which is not formulated in a higher-order predicate calculus (cf. Wang, 'Denumerable Bases of Formal Systems', p.76).

2   Cf. Strawson, *Individuals*, chs. 1 and 2.

3   We shall not consider the accounts of sense-perception which have been put forward in terms of 'sense data' since these have been successfully criticised by Austin (*Sense and Sensibilia*) and Sellars (*Science, Perception and Reality*).

4   Eddington, *The Nature of the Physical World*, pp.5-12.

5   Maxwell, 'The Ontological Status of Theoretical Entities'.

6   Čapek (*Philosophical Impact of Contemporary Physics*, ch. 5) adds motion to the list, which I think is correct, but this raises problems which it would not be in place to discuss here; so to keep the account simple I have excluded motion. We are only giving an example, not making a case study.

7   Cf. Mach, *Science of Mechanics*, p.577 ff.

8   Feyerabend, 'Realism and Instrumentalism'.

9   Nye, *Molecular Reality*, p.x.

10 A classic statement of this position is to be found in Carnap, *Foundations of Logic and Mathematics*. The most extreme variant of the doctrine is Bridgman's operationalist theory of meaning, on which the meaning of a theoretical term is synonymous with a corresponding set of operations (cf. Bridgman, *The Logic of Modern Physics*, ch. 1.)

11 The kind of ontological priority we are concerned with here is what can be called *strong* ontological priority. In this sense, 'X is ontologically prior to Y' means 'Y's are ultimately reducible to X's'. There is also a *weaker* sense of ontological priority, which seems to be what Strawson (cf. *Individuals*, Part 1) has

in mind when he uses this expression. In the weaker sense, '*X* is ontologically prior to *Y*' means 'it is a necessary condition of including *Y*'s in our ontology that we include *X*'s in it.'

12   Sellars, 'The Language of Theories', p.76.
13   Cf. Feyerabend, 'Explanation, Reduction and Empiricism'.
14   Sellars, *Science, Perception and Reality*, ch. 5.
15   See, for example, Feyerabend, *Against Method*, p.165 ff.
16   Cf. *ibid*, p.29 ff.
17   *Ibid,* p.73.
18   Austin, *Sense and Sensibilia*, p.76.
19   Sellars, 'Scientific Realism or Irenic Instrumentalism', p.172. The position of Feyerabend's which is being criticised here is his account of statements as 'behaviour orienting perceptual responses to sense-experiences'. Feyerabend characterises 'observation sentences' in terms of causal or behavioural responses to sensations, and they are distinguished from other kinds of statement 'by the psychological, or physiological, or physical circumstances of their production'. (Feyerabend, 'Problems of Empiricism', p.152.)
20   It is especially complex when we come to consider the question of how atheoretical statements can block the generation of explanations. A problem of this kind is dealt with in Bachelard, *La Formation de l'Esprit Scientifique*, but Bachelard's account is far from satisfactory (cf. Gaukroger, 'Bachelard and the Problem of Epistemological Analysis', pp.194-195, 227-233, 243-244).
21   Nagel, *The Structure of Science*, pp.146-151.
22   Cf. Haack, *Deviant Logic*, pp.133-5. As Haack points out, the basic problem is that Meinong's introduction of contradictory definite descriptions (such as 'the round square is surely as round as it is square'), if this is taken as a proposal for allowing the introduction of contradictory singular terms into a formal system, results in an inconsistent system. For example, using the usual rules of inference, inconsistency would result in the form of a theorem like:

$$F[(\imath x)Fx \ \& \sim Fx] \ \& \sim F[(\imath x)Fx \ \& \sim Fx].$$

The general problem of existence is complicated somewhat when we come to consider individuals in other possible worlds. Dana Scott ('Advice to Modal Logicians'), amongst others, has suggested the introduction of actual, possible and virtual individuals to deal with problems in modal logic. In Scott's notation, the domain *D* consists of actual and possible individuals but no virtual individuals. According to Scott, we can quantify over D; indeed, he asserts that D is 'fixed in advance'. For a discussion of some of the general philosophical issues involved here see Hintikka, 'The Semantics of Modal Notions and the Indeterminacy of Ontology'.

23 See above, footnote 11.

24 Mellor, 'Physics and Furniture', p.185.

25 Čapek, *op cit*, Part 1.

26 An idea of the complexities involved here can be gleaned from the fact that Čapek spends the first 100 pages of his *Philosophical Impact of Contemporary Physics* trying to sort out the ontology of a particular version of classical mechanics. His account is far from satisfactory in many respects (mainly because he attempts to treat classical mechanics in terms of the one ontology) but it is a serious attempt to investigate the structure of an ontology. It is worth comparing to another equally serious attempt of this kind: that given by Bachelard in his *Intuitions Atomistiques*.

CHAPTER 3

EXPLANATORY STRUCTURES

§1 Introduction

In the preliminary discussion of explanatory structures, I said that an explanatory structure consists of an ontology and a domain of evidence, and that these are linked together. With regard to the question of ontology, I have tried to establish two main points. First, I have argued that an ontology is a structured set of kinds of entity: it is 'structured' inasmuch as the kinds of entity which figure in the ontology bear definite relations to one another. Secondly, I have argued that an ontology need not reproduce any essential features of what we can observe.

The importance of ontological questions arises from the fact that it is primarily in terms of their ontologies that explanations are proposed in theoretical discourses. The fact that the ontology with which a theoretical discourse operates need not reproduce any essential features of observable things does not mean that explanations given in that discourse cannot explain what is observed: a molecular or subatomic account of repulsive forces may be put forward to explain the hardness of an object such as a table. The crucial point here is that an account of one kind of phenomenon — which figures in the domain of investigation — is given in terms of another kind of phenomenon — which figures in the ontology of the discourse in which the explanation is given. It is quite fortuitous that, in the example, the phenomenon which figures in the domain of investigation is observable and that which figures in the ontology is not observable. We might just as easily have explained something unobservable in terms of something observable, and so on.

With regard to the question of evidence, two main points have been established. First, I have argued that if something

[margin handwritten note: Cf. Hempel on internal principles & bridge principles, & the relns between observables & theoretical entities.]

61

is to count as an explanation — or, at least, if it is to count as a candidate for an explanation — then there must be evidence that would disconfirm (or tend to falsify) it. On the account of explanation being proposed, unless there could be evidence for and against a statement, that statement cannot be counted an explanation.[1] At a more general level, this is to say that discourses are the kinds of things which can generate explanations only if they operate with a domain of evidence. To say that discourses which operate with explanatory structures are the kinds of thing which can generate explanations, in principle, is not to say that such discourses can, as a matter of fact, generate explanations. For various reasons — which will be our concern in later chapters — they may not be able to do this. Secondly, we have seen that what could count as evidence for an explanation depends on the discourse in which that explanation is proposed. The discourse does not determine what the evidence for a particular explanation is, only what could count as evidence for that explanation.

The way in which the domain of evidence is circumscribed in any particular discourse is a problem that we shall have to deal with in some detail. We shall begin our account of this question with an examination of the connection between the entities in terms of which explanations are given and the circumstances under which those explanations can be shown to be sound or unsound. Because it is ultimately in terms of the ontology of a discourse that explanations are given, the ontology imposes important constraints on what counts as an explanation in that discourse. In this chapter, I want to try and specify the nature of these constraints, and in doing so I also want to give an account of the relation between what counts as an explanation in a discourse and what counts as evidence in that discourse.

§2.1   Domains of Evidence

The connection or 'link' between the ontology of a discourse and its evidential domain is a link between the structured set of entities in terms of which explanations are given and the set of situations which could count as evidential. The ontology of a discourse determines the kinds of entities in terms of which explanations can be given, but this is not to say that only those entities which figure in the ontology can be employed in the explanations of that discourse. Although they *ultimately* play this role, other kinds of entity can figure in explanations, but these other kinds of entity must be reducible, ultimately, to those kinds which *do* figure in the ontology. It may be thought that reducible entities are redundant. However, this is true only insofar as they are ultimately redundant: reference to them may nevertheless have to be made either in an area where a discourse is particularly underdeveloped, or when — for pragmatic reasons or for reasons of approximation — they serve a particular function adequately.

Insofar as it is ultimately the ontology of a discourse that determines which entities can, and which entities cannot, figure in explanations in that discourse, the ontology circumscribes what will and what will not be candidates for explanation. That is, it provides conditions for something's being of an explanatory kind. In molecular physics, for example, the statement:

(1) 'The hardness of tables is due to the repulsive forces between molecules'

is of the correct explanatory kind, whereas the statement:

(2) 'The hardness of tables is due to the innate power of matter to resist'

is not of the correct explanatory kind. The ontology of a discourse does not provide the necessary and sufficient conditions for adequate explanation, or even the necessary conditions for adequare explanation, but only *a* necessary condition for something's being counted as a candidate for an explanation in that discourse. Although this is not the only such necessary condition, it is the most important one for present purposes. Other necessary conditions include such things as the requirement that explanatory statements be well-formed. For example, the following statement does not account as a candidate for an explanation in molecular physics:

(3) 'Hardness, repulsive, table, molecule, force.'

This is a trivial example, of course, but in cases where the form of explanations is determined by the rules of syllogistic reasoning, or by algebraic rules, in the case of explanations which take the form of — or which make reference to — mathematical equations, requirements of this kind have an important function. Indeed, in the latter case, formal constraints which are rather more stringent than the requirement that explanatory statements be well-formed may be imposed. Considerations of formal analogy — and the increase in generality which may result — may require that equations of a particular kind be preferred. An example here would be the preference for second-order partial differentials in mechanics.

The ontology of a discourse imposes crucial constraints on the kind of explanations which can be given, but it does not determine whether explanations are correct. Statement (1) is of the correct kind in molecular physics, for example, but so too is the following statement:

(4) 'The hardness of tables is due to the attractive forces between molecules'.

A statement's being a candidate for an explanation is not the same as its being an explanation. (1) and (4) are both candidates for explanations in molecular physics, but if one of them is a (complete) explanation then the other cannot be. The problem we must now deal with is this: given that we have several explanations of the appropriate kind, how do we decide which of them, if any, is sound? In answering this question we must introduce considerations of evidence.

The domain of evidence specifies what could count as the relevant information in terms of which explanations are to be assessed: in short, what could count as evidence. This is not the same as saying that it determines what the evidence is, for it clearly cannot do this. In astronomy, for example, the domain of evidence may consist entirely of planetary motions, but the theory itself cannot dictate to the planets what paths they will take. The actual paths will be the evidence in terms of which explanatory accounts are to be assessed. What could count as evidence for a particular account, and what the evidence for that account is, are two different things. The first is determined by the discourse in which the account is put forward, whereas the second is quite independent of the discourse.

Properly included in any domain of evidence is a domain of *what actually does* count as evidence. The importance of distinguishing between what could count as evidence and what does count as evidence can best be illustrated by considering a hypothetical example. Let us suppose that we have two different accounts of the nature of the lunar surface. On the first account, the moon is perfectly spherical and has a perfectly smooth surface; on the second account the surface of the moon is craterous or mountainous. Now the first account, let us say, is allied to an epistemological theory of perception and a corresponding theory of optics in which the use of telescopes in extra-terrestrial observation is proscribed; the second account, on the other hand, is dependent on the use of telescopes. The difference here lies in the domains of evidence of the two accounts. The results of telescopic observation count as evidence in one account but not in the other; that is, these results figure in the domain of evidence of the second account but not in that of the first. This is not to say that there is no overlap between the two domains of evidence. There would be an overlap, for example, in a limiting case such as that where we actually travel to the moon and examine its surface. What distinguishes this limiting case is the fact that we do not have *access* to it; or, at least, we can posit that we do not have access to it in the situation mentioned (which is a *caricature* of an early seventeenth century debate). This leads us to the characteristic feature of that part of the domain of evidence which determines what *does* — as opposed to what *could* — count as evidence. What does count as evidence can be defined precisely as that part of the domain of evidence to which we have access: we can call it the 'accessible domain of evidence'.

In the case where we wish to compare two theories directly, the accessible domain of evidence is important. If we are dealing with sixteenth and seventeenth century debates on the nature of the lunar surface, for example, it is of no use to refer to the latest NASA findings. The results of a close-up observation of the moon's surface would figure in the domains of evidence of Peripatetic and Galilean theories, of course, but they would not figure in the *accessible* domains of evidence of these theories. Where accessible domains differ, direct comparison becomes problematic; where there is no overlap at all between accessible domains, direct comparison in terms of evidence becomes impossible. It is therefore of some importance that we be able to specify not only what the domain of evidence of a discourse is, but also what its accessible domain of evidence is. Further, discourses develop — they are not static — so we might well expect changes in domains of evidence, and changes in accessible domains of evidence. Changes in the former will generally have much more radical consequences than changes in the latter: variations in what could count as evidence for or against a theory are much more fundamental than variations in what actually does count as evidence.

Inasmuch as no appeal can be made to phenomena which fall outside the domain of evidence of a discourse, the domain of evidence will impose constraints on acceptable explanations in that discourse. These constraints are reciprocal. The kinds of explanation which are sought in a discourse will constrain the kinds of thing that can be appealed to as evidence; these constraints are open to revision but only under conditions that are operative in the discourse in question. When we come to the case studies in Part II, it will be an important part of our project to determine what constraints lead to the domain of evidence and the accessible domain of evidence taking the form they do in particular discourses, and to determine the conditions under which these can be revised. This will be done in order to provide us with a framework within which we can ask why the domain of evidence of a discourse is open to certain kinds of development and not others.

Let us now turn to a rather different question concerning access to the domain of evidence. In the hypothetical example of the lunar surface that we have just considered, the limiting case is one to which we have no access. A different kind of problem arises when the limiting case is the only one to which we have access, but where there is no direct connection between this case and the explanations produced in the discourse.

This can best be illustrated by briefly considering an

instance of the Law of Universal Gravitation: the case of bodies falling freely to the earth. This instance of the Law can be presented in the form of a statement about bodies falling in a vacuum: that is, a statement about the circumstances that hold when the only forces acting on the falling body are gravitational forces. Included in the domain of evidence here is the situation in which bodies fall in a vacuum, but we have no access to this situation. We do have access to the situation in which bodies fall in resisting media. In this case, we can say that there is no disagreement between the adherents to and the dissenters from the Law on the question of what situations we have access to. If we assume that there is agreement that the concept of a vacuum is not incoherent, then the new problem of access is this: how do we make the situation that we have access to part of the domain of evidence of our theory? After Galileo's pioneering work on the kinematics of falling bodies we have a relatively straightforward solution to this problem: we determine the nature and value of the parameters and variables which hold between a body falling in a vacuum and that same body falling in a particular medium. This procedure enables the construction of a conceptual link between two situations — one which figures in the explanation and one which figures in the domain of evidence. The concepts which form this link specify the parameters and variables that hold between the two situations — concepts such as 'mass', 'buoyancy effect of media', 'friction effect of media' and so on. The conceptual isolation of such variables allows the problem of their respective effects to be posed in such a way that these effects can then be subjected to experimental examination. This experimental examination can range from statistically corrected measurements to 'thought experiments' in which options can be rejected or suggested on non-quantitative grounds.

In this way, we can establish what is involved in the move from two 'mass points' separated by a certain distance in empty space, to two extended bodies (the earth and the falling body) separated by a resisting medium. The 'conceptual link' between the two situations is primarily that between the set of entities, and their possible relations, which figure in the ontology of the discourse (such as mass points in empty space, the relations between which obey the super-position principle) and the domain of evidence (extended bodies of non-uniform density in a resisting medium).

Now it is clear that if we are to have a workable theory, the domain of evidence for our proposed explanations must be such that we have access to it. A discourse which operates with a domain of evidence to which we have no access in principle cannot generate explanations. Further, we must be able to distinguish between evidence for and evidence against any purported explanations in an unambiguous fashion. The basis of a conceptualisation which is designed to achieve these two aims must ultimately lie in the kinds of entities which are invoked in explanation, for in explaining some-thing we are linking these kinds of entities to an evidential domain.

It might be argued that this is to confuse explanations with evidential support for, or 'testing' of, explanations, This is not the case however. I have argued that it is part of the concept of explanation that in explaining something an evi-dential domain be (or could be) invoked: statements which give an account of something in a situation where questions of evidence are not appropriate cannot be explanatory since they are atheoretical in the sense defined in the last chapter. Moreover, in invoking an evidential domain, we must be able to stipulate what access we have to that domain, since it is only when we have done this that we can make evidence available.

To sum up, the linking of the ontology of a discourse to its evidential domain is achieved by a system of concepts peculiar to that discourse. This system of concepts is *peculiar* to the discourse insofar as it has its roots in the ontology of that discourse. The kinds of entity which figure in the ontology must be related to one another, but when these kinds of entity figure in the explanations of the discourse, they must also be related to those concepts which are used to demarcate the domain of evidence. A correlative require-ment is that we be able to specify the conditions under which access to (at least part of) this domain — and the information it contains — is possible. This means that the conceptual structure of a discourse must be capable not only of linking the situations described in the laws or principles which form the basis for particular explanations to evidential situations, but also that it be capable of specifying the conditions under which we have access to the evidential situations.

§2.2 Domains of Evidence and Concept Formation

We must now turn to a related issue which, although at first sight it seems to have little to do with the question of evidence, is, I shall argue, simply part of the more general problem of linking evidential domains to ontologies. This is the problem of construction of concepts.

If we are to be able to give an explanatory account of a particular area of investigation, this must be conceptualised in some way. One of the main ways in which theoretical discourses are constituted in the first place is by such a conceptualisation, and the kind of explanatory account that we wish to give of an area of investigation will clearly have a bearing on the way in which it is to be conceptualised. The problem of conceptualisation is a problem concerning the kinds of concepts required if a particular area is to be investigated in such a way as to generate explanations. This, in turn, raises the problem of the means by which concepts are generated. There are two main points that I wish to make about this latter problem. First, it is not a psychological problem: we are dealing with concepts not *qua* 'ideas in people's heads' but *qua* constituents of theoretical discourses. Secondly, the 'means' by which concepts are formulated are variable, depending on the explanatory structure of the discourse in question, and on the kinds of problems which these concepts are formulated to deal with.

The problems of concept formation that we are interested in are not ones which concern the way in which an individual practitioner of a discourse produces new concepts or a new theory. On this kind of problem I think Popper is probably correct when he claims that the 'procedure of finding a hypothesis cannot be rationally reconstructed'.[2] What I shall be concerned with is the way in which a discourse constrains the procedures by which new concepts and theories can be produced. These constraints can, and indeed *must*, be 'rationally reconstructed'. The procedures to which they give rise — whether these be characterised in terms of induction or whatever — are of secondary importance given our main concern. If we are interested in analysing the conceptual difference between Aristotelian and Galilean mechanics, for example, the issues do not turn on the question of whether induction is used or not,[3] but rather on the constraints that are operative in the production and assessment of new theories in these discourses. After all, we cannot invoke induction to *explain* how we come by concepts or theories since we require concepts or theories in the first place to be able to perform an induction. At the *explanatory* level, induction is redundant. Theories determine what kinds of things should be examined, the conditions under which they should be examined, and the conclusions that can legitimately be drawn from these examinations. This means that if we want to give an account of why certain kinds of conclusions are drawn from certain kinds of investigation we must examine the theories involved; a study of 'induction', in the abstract, would tell us nothing about these issues. It is the theory that determines what counts as the relevant evidence, how we come by this evidence and whether the evidence is adequate. In considering the question of adequacy, such things as probability theory may well be invoked, but whether and how they are invoked is something which depends to a large extent on the original theory.

The problem of the construction of concepts is one which can only be resolved by asking what particular conceptualisations are designed to achieve, and what kinds of investigations would be required in order that such conceptualisations might be shown to be sound. Given a certain area of investigation, and given an account of what an explanation is, concepts are required which will enable us to pose problems about that area of investigation which will result in solutions which have explanatory value. This means that particular kinds of problems must be posed and not others; moreover, these problems must be posed in a particular way and not others so that they are amenable to particular kinds of solution procedures. After all, theories — and theoretical discourses — arise in the first place when particular problems are posed in particular ways with a view to obtaining a particular kind of solution (*viz*, one with some explanatory power). We shall examine the problem of solution procedures when we come to consider the question of proof in the next section, but it will be clear from what has already been said that since the explanatory structure of a discourse imposes constraints on what is to count as an explanation in that discourse, it must also impose constraints on the way in which problems can be posed. This means that it imposes constraints on the way in which problems about particular areas are to be conceptualised.

The variability in the way in which problems are to be conceptualised involves a requisite variability in the way in which the necessary concepts are come by (or 'constructed' or 'developed'). In any particular discourse, some things can count as a source of concepts and others cannot. Everyday experience is particularly important as a source of concepts in Aristotelian physics, for example, but this is not the case in physics from Galileo onwards. There, the kinds of conceptualisation of problems which can be achieved on the basis of a careful consideration of everyday experience no longer serve as an adequate basis for generating explanations.

Given a particular discourse, it is only those problems which can be posed in a particular way whose solutions can count as explanations. In order for problems to be posed in this way they must be conceptualised in a particular fashion. What concepts are required will clearly depend, to some extent, on the domain of investigation. More than this must be involved, however, for there are plenty of examples of cases of discourses with the same or overlapping domains of investigation (referentially speaking), but which operate with quite different concepts. The way in which local motion is dealt with in Aristotelian and classical mechanics is quite different, for example. The primary factors in determining how problems about a certain domain of investigation are to be conceptualised are (1) the kinds of entities which are going to be invoked in the explanation, and (2) the kinds of phenomena which are going to be introduced as evidence for those explanations. It is because of this that questions of evidence play a crucial role not only in the 'testing' of explanations that have already been produced, but also in the formulation of the very specific problems whose answers can have explanatory value in that discourse: where one of the criteria by which we decide whether an account has explanatory value, for any particular discourse, is that the account should be assessable in terms of evidence. In constructing a problem within a discourse, the concepts in terms of which that problem is posed must be such that the possible solutions to the problem are assessable in terms of the domain of evidence of the discourse in question. This means that the domain of evidence of a discourse plays at least as important a part in the construction of problems as it does in the assessment of their solutions. Moreover, in constraining the kinds of concepts which are available in a discourse, the domain of evidence thereby constrains which new theories in that discourse can be considered viable candidates for explanatory accounts. We may well not be able to give an account of why the individual practitioner of a particular discourse produces particular new theories, but by examining the explanatory structure of a discourse we can specify whether any new theory can be considered a viable explanatory account within that discourse, and we can also specify what kinds of constraints are operative in the formulation of new concepts and new theories.

§3 Proof Structure

Up to now, I have argued that explanatory structures consist of an ontology linked to a domain of evidence, and that one central way in which this link is effected is via a system of concepts which serves to relate the kind of situations which can be specified in terms of the ontology of the discourse to situations that figure in the domain of evidence and to which we have access. Because this system of concepts relates a particular ontology to a particular domain of evidence, it is *peculiar* to the discourse. However, this is not the only system of concepts which a discourse requires, and we must now consider the role of systems of concepts which are not peculiar to the discourses in which they operate.

The most important non-peculiar system of concepts with which a discourse operates is that system which provides the constraints on consequence and derivation relations, and we shall call this non-peculiar system of concepts a *proof structure* from now on, reserving the term 'system of concepts' exclusively for the system of concepts which links the ontology of a discourse to its evidential domain. This system of concepts is a *system* inasmuch as the concepts are related in a definite fashion. It is in connection with the problem of this system of concepts *qua system* that the issue of proof structure arises. There are three topics here which we shall consider briefly. The first concerns the nature and function of proof structures. Secondly, there is the question of how proof structures can differ, and why this difference arises. Thirdly, there is the problem of the effect of different proof structures: what constraints do they impose on other parts of the explanatory structure, and what constraints do other parts of the explanatory structure impose on them? Now the range of problems which these issues raise is vast, and I shall not pretend to give a full account of them here. We shall concentrate on the kinds of problem that will arise in Part II, but other issues will also be mentioned.

At the most general level, the problem is that of understanding the function of logical and mathematical operations in different discourses. This problem is a reciprocal one: ideally, we should be able to understand how logical and mathematical concepts and techniques function, for any particular case, outside pure logic and mathematics; also, we should be able to understand how concepts which are not physical concepts, for example, function in particular physical discourses. It is the second of these issues to which our attention will be directed primarily, since it is here that the main problems arise when we come to consider Aristotelian and Galilean mechanics. It is a commonplace that Aristotle proscribes the use of mathematical theorems in dealing with physical problems, and that mathematical procedures have a fundamental role to play in classical mechanics. Such a situation clearly involves a difference in what counts as an explanation — and hence this is an issue which we should be able to deal with by an analysis of the explanatory structures involved. The difference, however, is of a kind that we have not yet considered. It is important therefore that we at least be able to identify the issues that are involved here.

These issues centre around the problem of proof. The *proof structure* of a discourse is whatever provides the constraints on the formal relations between the concepts of a discourse, and hence on the statements produced in that discourse. In the simplest case, the only requirement may be that the relations between statements — together with the relations between components of 'compound' (i.e. conjunctive or disjunctive) statements — obey some fundamental logical theorem, such as the law of the excluded middle. In more complex cases, such as those in which we wish to calculate, other requirements also come into play. In calculation, for example, the standard procedure is to associate particular entities/sets of entities, or relations between entities/sets of entities, with quantities or magnitudes, and then to construe the relations in such a way that they are isomorphic with the relations between numbers (or whatever kind of mathematical entity one is operating with). Here, the kinds of operations which can be performed become subject to the laws of mathematics. Moreover, the association of entities with magnitudes presupposes a theory of measurement. Such a theory is rarely explicit in any pre-quantum mechanical theories, and indeed the necessity for a theory of measurement was probably not explicitly realised before the work of Herz and Mach. It is only by means of a theory of measurement that calculation and measurement procedures can be related, and such a theory may, in some cases, involve problems with radical logical consequences: it has been argued, for example, by Finkelstein, Putnam and others that measurement problems in quantum mechanics may require the adoption of a logic in which the distributive law does not hold.[4]

The proof structure of a discourse, in constraining the

relations which can hold between statements, thereby constrains what counts as a valid proof (or demonstration or derivation or even, in the limiting case, statistical inference) in that discourse. That is, it constrains the kinds of concepts and techniques which can be used in the various proofs and demonstrations with which a discourse operates. In doing this, it plays an important role in determining not only how problems are to be resolved, but also how they are to be posed. Mechanical problems in which calculations are involved, for example, cannot be posed, in general, unless we operate with quantitative concepts and a symbolic concept of number. The importance of the proof structure of a discourse derives not so much from the constraints it imposes on the resolution of problems but rather from the constraints it imposes on the mode of presentation of problems. The latter is much more fundamental, for the way in which problems can be posed or presented in a discourse clearly has an important bearing on the kinds of explanations which can be sought in that discourse. For example, not all discourses operate with procedures for calculation. This is not simply because not all discourses can realise the conditions under which calculation is possible. In some cases, the procedure of posing physical problems in a mathematical form is simply not a procedure which is commensurate with the requirements for explanation in those discourses. In Aristotelian physics, for example, a complex system of 'natural' classification must be adhered to if explanations of the required kind are to be generated. As I hope to show in the next chapter, the proof requirements of this physics include a condition of homogeneity between the premisses and the conclusion of a syllogism which the use of mathematical theorems and concepts in physical explanation does not satisfy. The conception of explanation which is at work here precludes the use of mathematical calculations in physics on the grounds that they are simply not appropriate.

One important feature of the distinction between a proof structure and a system of concepts is that it enables us to distinguish clearly between two different ways in which non-peculiar concepts can be introduced into a discourse. The first kind of case is that where concepts which are initially non-peculiar are transformed into the peculiar concepts of the discourse by being introduced on a reductionist basis. Here, the new concepts are introduced directly into the conceptual system of the discourse, not into its proof structure. For example, in the case of the introduction of physical concepts into chemistry, or chemical concepts into biology, the latter become (at least partially) subsumed under the former; most importantly, this subsumption can be an ontological subsumption so that the ontology of a chemical discourse, for example, is reduced to that of a physical discourse. The chemical discourse then becomes a branch of the physical discourse, distinguishable from other branches by its domain of investigation.

The introduction of mathematical procedures into a physical discourse, on the other hand, is quite different from this. No reduction is involved in this case since mathematics operates at the level of the proof structure: mathematical concepts are not introduced directly into the conceptual system of the physical discourse. The introduction of new concepts into the proof structure of a discourse may, of course, change the constraints which are operative in the system of concepts in a discourse, and this in turn may call for a revision of the conceptual structure. However, since the concepts which figure in the proof structure and those which figure in the conceptual system have different functions they cannot replace one another: what can happen is that changes in one can call for changes in the other.

Finally, it should be remarked that the relations which hold between an ontology, a domain of evidence, a system of concepts and a proof structure can only be specified as *possible* relations at the level we have been discussing these. What actual effects one has on the other is something which depends on the discourse in question, and on what one wants to do in this discourse. Problems of proof and formal presentation may require the introduction of new kinds of entity into the ontology of a discourse, or they may require that the kinds of entity which already figure in an ontology be related in a different way. Mathematical constraints on the (algebraic) formalisation of classical mechanics, for example, may require that gravitational potentials be introduced into equations, and this in turn may cause serious problems of ontological commitment. On the other hand, certain kinds of formalisation may be proscribed on the basis of ontological considerations.

§4 The Constraints on Explanation

In the last three chapters, and particularly in the present chapter, I have been concerned to identify the main kinds of factor which can go towards determining what counts as an explanation, and to relate these to the procedures which can be used to generate explanations. I have not tried to provide a general model of explanatory structures which can then be 'applied' to particular cases; rather, an attempt has been made to outline the general kinds of question which must be posed if we are to be able to analyse and assess the explanations which are produced in a particular discourse. At the abstract level, we can specify that ontologies, evidential domains, conceptual systems and proof structures provide important constraints on explanation in a discourse. This specification is 'abstract' in the sense that it can be established in a formal manner. I have argued that in explaining something in a theoretical discourse we operate with an ontology, and that the explanations given in terms of that ontology must be related to a set of phenomena which could count as evidence for these explanations; I have also argued that this requires that there be a system of concepts linking the ontology to the domain of evidence, and that the arguments and operations in which these concepts can figure are governed by rules determining what kinds of inference are valid and under what conditions. I have described these features of an explanatory structure in terms of the 'constraints' that they impose upon explanation. Explanations which were not constrained in some way would be vacuous, since anything would count as an explanation. It must be remembered, however, that these constraints do not operate in isolation from one another: they are interconnected. The precise way in which the constraints operate depends on the discourse in question, and upon the kinds of problems encountered in that discourse at any stage of its development. They are open to revision, but such revision is limited by the effects on other parts of the discourse: one cannot simply introduce mathematical proofs into Aristotelian physics, for example, without calling into question the whole enterprise of attempting to discover essential principles from which the properties of physical bodies can then be derived (as we shall see very shortly).

The kinds of questions which, I have argued, must be posed in the examination of explanatory structures, are designed to enable us to locate explanatory problems at their source and thereby to allow us to distinguish different types of problems which, although they may all result in the same phenomenon — the inability to produce explanations — have quite different causes. In attempting to isolate and analyse explanatory problems in two discourses in detail in the next three chapters, we shall find that the ways in which the constraints, that an explanatory structure imposes upon explanation, operate are very complex. We shall also find that the priorities which are given to different kinds of constraint vary considerably, and that this variation is determined by factors which cannot be specified in advance. Most importantly, however, it is the constraints which are operative in a discourse that determine what is to count as an explanation in that discourse, and it is only by examining these constraints that we can determine why particular kinds of explanation are sought in particular cases, and why particular kinds of explanatory difficulty are encountered.

Up to now, I have attempted to present a general account of the kind of issues which must be raised in isolating, analysing and assessing explanatory structures. We are now in a position to put this account into practice, and in doing so to develop and refine it: and ultimately, of course, to test its usefulness. In chapter 7 we shall return to the general issues that I have raised here and discuss them in the light of findings of Part II.

Notes: Chapter 3

1 The conjunctions or disjunctions of statements (or derivations from these) may fulfil this requirement in cases where the individual statements taken by themselves do not. These conjunctions or disjunctions (or their derivations) may then be taken as individual statements fulfilling the requirement. Not all statements in a discourse *will* usually be tested evidentially and Lakatos is surely right in arguing that parts of discourses must be kept free from criticism if these discourses are to develop — particularly in their early stages.

2 Popper, *Logic of Scientific Discovery*, p.315.

3 This applies equally to those — such as Feyerabend (*Against Methods,* chs. 2 and 3) — who consider that the characteristic feature of Galilean mechanics is that it proceeds by counter-induction. There are plenty of instances of both induction and counterinduction in Galileo: and in Aristotle for that matter (the theory that continued projectile motion requires that the air exert a force on the projectile is arrived at counterinductively for example).

4 Cf. Finkelstein, 'Matter, Space and Logic', and Putnam, *Philosophical Papers*, I, Essay 10. Putnam argues that 'the only laws of classical logic that are given up in quantum logic are distributive laws, e.g. p.(q v r) = p.q v q.r; and every single anomaly vanishes once we give these up' (*ibid*, p.184). That there can be no *a priori* objection to non-classical logics has been argued in Haack, *Deviant Logic*. Haack, however, opts for a global reform of logic which is, I think, seriously wrong. One of her main objections to local logics is that there must be some weakest logic which holds in all areas (p.166). But even if we accept that there may be some very weak minimal logic which, when supplemented by the appropriate theorems, can be made to yield classical logic, intuitionist logic and so on, this would not be a global reform of logic in any normal sense since the issue of local or global reform hinges on the question of whether *classical* logic holds universally, and not on the question of whether there is some minimal logic which holds universally.

PART II

CHAPTER 4

PHYSICAL EXPLANATION AS SYLLOGISTIC DEMONSTRATION : I

§1   The Object of Explanation

FOR Aristotle, 'science' or 'knowledge' (*epistēmē*) comprises three factors. First, there is that about which (*peri ho*) conclusions are established; second, there are whatever (*ha*) are established as conclusions; third, there is that from which (*ex hōn*) the conclusions are established. That about which conclusions are established is always a particular kind of subject matter for each particular science. The aim of scientific enquiry is to determine what kind of thing the subject matter is — by establishing its essential properties and 'causes' (*aitia*) — and to state what it is — to formulate its definition. Language (*logos*) is the instrument (*organon*) by which *epistēmē* is formulated and expressed. That from which conclusions are demonstrated are *archai* or 'principles'.

There are *archai* which are peculiar to particular sciences, but there are also general *archai* which are common to all sciences. The common *archai* are those of demonstration (*apodeixis*) itself. Demonstration can occur only in language, but language must be used in a certain way — in a syllogistic form — if it is to be truly demonstrative. Moreover, of the three most general kinds of syllogism — dialectical, eristic and demonstrative — only the last produces *epistēmē*. Since the syllogism is a purely formal mode of reasoning,[1] these three kinds of syllogism are formally equivalent. The difference between them lies in the nature of their respective premises. In the dialectical syllogism, the premises are authoritative opinions (*endoxa*).[2] In the eristic syllogism, the premises are opinions (*endoxa*) which seem to be generally accepted but which argument (usually *reductio ad absurdum*)

83

shows are not, or could not really be, accepted.[3] We shall
discuss the premises of the demonstrative syllogism later;
for the moment we can call them the 'phenomena'
(*phainomena, erga, huparchonta*).

Any *scientific* explanation must be syllogistic in form but
it must also be *demonstrative*. An examination of the
conditions under which a syllogism can be said to be
demonstrative will enable us to isolate the conditions under
which one can be said to have given an explanation of a
phenomenon. The demonstrative syllogism is that form of
syllogism — strictly speaking the *only* form of syllogism —
appropriate and adequate to the theoretical sciences:

> 'By demonstration I mean a scientific deduction; and by scientific I
> mean one in virtue of which, by having it, we understand something.
> If, then, understanding is as we posited, it is necessary for
> demonstrative understanding in particular to depend on things
> which are true and primitive and immediate and better known than
> and prior to and explanatory of the conclusions, for in this way the
> principles will also be appropriate to [i.e. in the same genus as] what
> is being proved. For there will be deduction even without these, but
> there will not be demonstration; for it will not produce
> understanding.'[4]

That upon which demonstrative understanding depends must
be true because, although true conclusions can be drawn
from false premises, these are true 'only in respect to the
fact, not to the reason'.[5] It must be primitive and immediate,
otherwise it would have to be demonstrated, and if all
knowledge were demonstrative we would be committed to an
infinite regress. The contention that the conclusion must be
related to the premises in an 'explanatory' fashion — or, as
it is translated in the Ross edition, 'as effect to cause' — is
the most central feature of Aristotle's account of demonstration: 'We possess scientific knowledge of a thing only when
we know its reason [or "cause" or "explanation"]'.[6] This
being the case, the premises must be prior to the conclusion,
since because they are the 'cause' of our knowledge we must
know them better than their consequences, 'precisely because
our knowledge of the latter is due to our knowledge of the
premisses'.[7]

What we are seeking in the demonstrative syllogism is
knowledge of things that cannot be other than they are.[8] We

are not seeking knowledge of particular sensible things as such because these can be other than they are. Scientific knowledge of nature (*phusis*), for example, is not simply knowledge of physical things for Aristotle, but knowledge of the nature and interrelationships of the causes (*aitia*) of things, and of *phusis* known through its *aitia*. Knowledge lies in the demonstration of connections that are eternal, necessary and 'commensurately universal'. Since knowledge is of what cannot be otherwise the 'causal' connection between the premisses and the conclusion of a demonstration is an eternal one. With regard to necessity, 'all attributes which . . . are essential either in the sense that their subjects are contained in them, or in the sense that they are contained in their subjects, are necessarily as well as consequently connected with their subject.'[9]

There is a commensurately universal relation between a subject and an attribute when the attribute 'belongs to every instance of the subject, and to every instance essentially and as such; from whence it follows that all commensurate universals inhere *necessarily* in their subjects.'[10] This commensurately universal relation between subject and attribute is given, in the demonstrative syllogism, in the relation between the minor term (the subject term of the conclusion) and the major term (the predicate term of the conclusion) which, through the middle term, serves to connect the minor and major premisses. It is the middle term which provides the connecting link between conclusion and premisses, since it provides the 'reason why' (*to dioti*). In actual scientific work, the aim is to find such middle terms and in doing so to trace an intelligible structure between things, thereby exhibiting the general intelligible structure of phenomena.

These middle terms are the *archai* of the particular sciences. As such, they determine both the subject matter of the particular science, its mode of proof and, to a considerable extent, the domain of evidence for that particular science. Before we can discuss these issues in detail, there are two questions we must answer. They are: (1) What conditions must something fulfil if it is to be an *archē*? and (2) How do we come by *archai*?

The object of scientific enquiry is a *definition*, a statement of the 'nature' (*logos*) of something. In the *Metaphysics*, Aristotle argues that it is substance (*ousia*) which is the object of definition. In the course of the discussion of what counts as substance[11] he lists four possibilities: essence (*einai*), universal (*katholou*), genus (*genos*), and subject (*hupokeimenon*). Taking substance as the ultimate subject, Aristotle asks whether this is to be identified with matter, form or the combination of the two.[12] Since matter without form is not determinate, and since the combination of matter and form is logically posterior to matter and form taken separately, neither of these will do, and we are left with form (*eidos*). This is identified with essence (*to ti en einai*), which is in turn identified with 'primary substance' (*prote ousia*) at 1032b1-2.[13]

Universals, however, cannot be identified, as such, with substances. Aristotle is quite explicit in ch 13 of Book Z of the *Metaphysics* that no universal term can be the name of a substance, because the substance of a thing is peculiar (*idios*) to it whereas that which is predicated universally (*legetai katholou*) can be common to many things.[14] In the case of living beings, this is not too much of a problem since, as Albritton[15] has shown, there is plenty of evidence (for example, in *De Anima* and in *Metaphysics* ∧ and M) that Aristotle considered that there is a form peculiar to each member of a species; furthermore, such a doctrine would be required to make sense of the claim that the form of animate things is their *psuche*. Woods[16] has argued that this doctrine in fact covers all beings — animate and inanimate. This is important because it has a crucial bearing on how we conceive the process whereby we 'discover' *archai*.

Individual substances are individual in virtue of their forms. The individuals of a particular species are not individuated on the basis of a predicate that applies to them; rather, this particular predicate applies to them because they already possess a certain form. Because species *determine* their principles of differentiation — rather than being *determined by* their principles of differentiation — they can be conceived as self-differentiating.[17] The claim that the species *form* cannot be predicated of a plurality of objects is really a claim about logical priority — the possession of a particular form by individual substances is logically prior to the differentiation of such objects into a plurality. We can state this briefly by saying that species form *natural kinds*. The demonstrative syllogism exhibits these kinds. In stating the definition of a thing its position in the general structure of kinds is indicated. On this basis, the essential properties of things — those properties which follow from the kind of thing something is [18] — can be exhibited. This would be a universal and necessary demonstration of the kind required. It is the 'nature' (*logos*) or 'cause' (*aition*) of the properties that is expressed in the definition of the kind of thing in question.

Now *aitia* are usually identified with *archai* in Aristotle's work, and in seeking the *archai* of a particular science we are seeking the *aitia* expressed in the definition of the subject matter of that science, so before discussing this procedure it would be of benefit to determine exactly what *aitia* are. As Wieland[19] has shown, the concept of *aition* is quite different from the modern concept of cause, or the Medieval concept of *causa*. What is 'caused' is not a change itself, but the result of this change. Aristotle does not speak of 'effects' but, rather, he tends to use circumlocutions such as 'that of which the causes are causes'.[20] This is important because, in general, causes and caused are not at the same conceptual level: in examining something for its causes one is seeking principles (*archai*) of things, whereas in dealing with what is caused one is dealing *primarily* with *things*. Unless we bear this in mind, much of what Aristotle writes about motion (*kinēsis*), in particular, will remain completely obscure. We cannot simply lift the concepts of cause and effect out of classical mechanics and read them into pre-classical mechanics.[21] An example of the misunderstanding which would result can be illustrated by the following remark from an otherwise excellent text on Aristotelian mechanics:

> 'Seeking a principle, Aristotle finds a quality of substance . . . [This results in] a confusion between movement produced and the production of movement.'[22]

Aristotle's concept of *aition* is not an incidental feature of his account. After all, since it is *aitia* that we are seeking in demonstration, when we have given the *aitia* of the phenomena we have given an explanation of them, and this is the whole point of scientific enquiry. From what has already been said, it will be clear that *aitia* are not abstractions, at least in any usual sense of this word. Further, Aristotle makes no attempt to deduce the four kinds of *aitia* logically.[23] Rather, he simply introduces them on the basis that the investigation of sensible change brings to light these four *aitia* and no others. This peculiar procedure requires some clarification.

In Book Δ of the *Metaphysics* we are told that causes are 'said in many ways', that there may be several causes of the same thing, and that these several causes are not causes 'in any accidental sense'.[24] The things which are 'said in many ways' have an identical name, but they have different definitions (*logoi*). Note that what is said to be equivocal here is the *thing* and not the *name*. Philologically, this is made possible by Aristotle's conversion of functional concepts and attributes into subjects. This philological innovation allows the analysis of the structural components of everyday speech by converting them into subjects.[25] This is important because of the relation between subjects, at the grammatical level, and things, at the ontological level. Although Aristotle usually uses the neuter singular or plural of the adjectival form — for example, 'the prior' rather than 'priority' — he often passes from one to the other without change of meaning. A parallel usage in English would be that of 'relation' and 'relative' in the sense of a person related. 'Relation' here denotes a *thing* (*viz*, a person). Indeed, for Aristotle, it is primarily *things* which are equivocal and not concepts or words — which simply signify things. Owens has argued that the equivocality of things exists *in rerum natura* for Aristotle:

'The "obscurities in the nature of concepts" is accordingly a misleading formulation of a real Aristotelian problem. The true difficulties — and Aristotle is keenly aware of them — lie in equivocals. These are primarily *things*. Things are in some ways the same, in some ways different. If a different term were used every time the definition differed, all danger would vanish. So too would be lost in the expression the unity and interrelations that groups of things have among themselves . . . To Aristotle, as to Plato, things appear from their very nature both same and different. Aristotle refuses to employ a different word for each concept and definition, predetermining each term to a definite technical meaning. He uses rather a method of speaking that will allow the sameness of things to remain manifest, while being careful to attend to their differences. *Things* are that way. Aristotle tries to regard them as such.'[26]

This claim is too strong. It is true that definition arises from the thing itself and not from the name of the thing, but for Aristotle names refer to things *insofar as they are known*.[27] It is simply inappropriate to consider things *in rerum natura* as being either equivocal or univocal. It is insofar as they are known and *named* that they are equivocal (or univocal). Equivocality is a feature of things as these are signified in language. For Aristotle, this is neither a failing nor an advantage of language. It is because of this that univocal concepts are not a *general* desideratum: equivocal concepts are necessary because they alone can allow us to mirror the differences between things. Indeed, in the final analysis (but with one proviso), it does not really matter whether it is things *in rerum natura* or things as they are known which are equivocal. At the level of science (*epistēmē*) we are interested in things as they are known, and the two versions of equivocality coincide. There would be an effective difference between the two kinds of equivocality only if (and this is the proviso) we were prepared to make an *ontological* distinction between realms of being and knowing, but Aristotle does not make a distinction of this kind. This *ontological* distinction is one peculiar to the Scholastics, yet Owens — like most Thomists — ascribes it to Aristotle himself.

The doctrine of equivocality is intimately connected with what I wish to argue are the two central features of Aristotle's account of *epistēmē*: the reliance on sense experience and the reliance on analysis of forms of speech. The former, which we shall examine later, serves to delimit the domain of *evidence* for any particular science. The latter serves to delimit the *ontologies* of the particular sciences. In order that it should do this, Aristotle finds it necessary to extend language in a new and extreme fashion. In particular, we could note the Aristotelian categories, where interrogative pronouns are turned into subjects: for example, *to poson*, quantity, or literally 'the how much?'.[28] Having done this, Aristotle never has to leave the analysis of forms of speech, and the doctrine of *archai* in the *Physics*, in particular, never leaves this realm. This is not to say that an analysis of this kinds is an end in itself for Aristotle: it is not. Rather, such an analysis is directed towards *things*, these things being mirrored in speech. Randall, when he discusses the demonstrative syllogism under the heading of 'Science as Right Talking'[29] puts his finger on a very important aspect of Aristotle's project, so long as we bear in mind that the criteria of 'rightness' here are not formal or linguistic but material, in that they relate to sensible things.

Language, if used correctly, mirrors things. These things are given in sense experience. They are not ultimately differentiated by some perceptual or conceptual structuring; rather, they are self-differentiating. It is the function of language to articulate — by use of the syllogism — the principles of this differentiation. Sense experience can be used to determine what exists, but it cannot, by itself, determine why what is is what it is. For this we need *archai* or *aitia*, and we must now turn to the question of how we come by these.

§2   The Link between Evidence and Ontology

The crucial passage on how we come by *archai* or *aitia* occurs at the end of the *Posterior Analytics*. It reads as follows:

> 'We conclude that these states of knowledge [*viz*, practical and scientific knowledge] neither belong in us in a determinate form [that is, "are innately present in us in an actualised state"], nor come about from other states that are more cognitive, but from sense perception. It is like a rout in a battle stopped first by one man making a stand and then another, until the original order [*archēn*] has been restored. The mind [*nous*] is such as to be capable of undergoing this. What we have just said, albeit rather unclearly, let us now repeat. When one of the undifferentiated things makes a stand, there is a primitive universal in the mind (for though one perceives the particular, perception is of the universal — e.g. of man but not of Callias the man); again a stand is made in these, until what has no parts and is universal stands — e.g. *such and such* an animal stands,

until animal does, and in this a stand is made in the same way. Thus it is clear that it is necessary for us to become familiar with the primitives by induction [*epagōgē*], for perception too instills the universal in this way.'[30]

The three main topics here are the role of sense perception, the role of 'induction', and the role of *nous*. We shall examine these in turn.

What is perceived in sense perception is the *phainomenon*, and it is from the *phainomena* that *archai* are derived:

'It is the business of experience to give the principles which belong to each subject. I mean, for example, that astronomical experience supplies the principles of astronomical science: for once the *phainomena* were adequately apprehended, the demonstrations of astronomy were discovered.'[31]

The *phainomena* here clearly include 'empirical observations'. As far as the meteorological, biological and astronomical works are concerned, 'empirical observations' would be reasonably accurate as an interpretation of the meaning of the word. The great stress that Aristotle lays on sense perception also suggests such an interpretation:

'All men desire by nature to know. An indication of this is the delight we take in our senses; for even apart from their usefulness they are loved for themselves; and above all others the sense of sight. For not only with a view to action, but even when we are not going to do anything, we prefer seeing (one might say) to everything else. The reason is that this, most of all the senses, makes us know and brings to light many differences between things.'[32]

The idea that knowledge is based primarily on sense perception is put forward in a more developed way in the *Posterior Analytics*. We are told there that the persistence of sense impressions leads to memory, which in turn leads to 'experience' [*empeiria*] which is the basis of empirical knowledge in that it provides one with knowledge 'that a thing is so' (the fact) but not yet 'why a thing is so' (the reasoned fact).[33] Finally in *Metaphysics*, Γ, 5, Aristotle argues that what sense perception tells us about what we experience is true (providing we distinguish between the physical objects and the 'affectations' that they produce.)

The situation is not quite so straightforward as it appears however. It is clear from passages in other texts that *phainomena* does not refer *simply* to 'empirical observations'. In the discussion of *acrasia* in the *Nichomachean*

Ethics,[34] for example, the word *phainomena* refers to what is usually said and believed, and also to what is held by the wise and by the ancients.[35] In practice, this conception of *phainomena* also functions outside the texts devoted to practical reasoning. In the *Physics*, for example, the starting point is the *endoxa* of Aristotle's predecessors. Although the examination of these *endoxa* does not and cannot take the form of demonstrative reasoning, it takes the form of dialectical reasoning which can in some sense be called 'scientific' in that dialectic has a 'use in relation to the ultimate bases of the *archai* used in the several sciences'; furthermore, 'dialectic is a process of criticism wherein lies the path to the *archai* of all enquiries'.[36] Dialectic is used here as a means of criticising proposed *archai* in terms of their consequences.[37] The usual mode of criticism is the *reductio ad absurdum*, but equally important is the failure of *archai* to agree with sense perception.

One finds a fundamental and exclusive reliance on sense perception in Aristotle's work. This reliance is exclusive inasmuch as although our starting point in producing scientific knowledge must be the *phainomena*, the *phainomena* include the teachings of authorities and what is commonly held, etc, only insofar as the consequences of these are in agreement with what is perceived. Much is often made of the comparatively wide extension of the term *phainomena*, but we must remember that the Paramenidean denial of change, Atomist theories, and so on, were commonly held and authoritative views. Nevertheless, Aristotle does not grant the same status to these as he does to sense perception: indeed, one of the reasons he denies their truth is that their consequences do not agree with sense perception.

The idea that perception is the arbiter of scientific theories is supported by Aristotle's account of this in the *De Anima*. In sense perception, the sense organ receives the form of the object perceived without its matter; the object actualises a potentiality that the organ has for receiving the forms of objects, so that the sense organ becomes what the object is. Before perception, the sense organ is potentially what the object is actually, and this potentiality is made actual in perception. In this way, 'actual knowledge' is identical with what is known. Knowing the nature (*phusis*) of something is the same as having that nature in the intellect. One actually has the *form* of what is known in the intellect but not, of course, the matter. The efficient cause of perception is the motion imparted to the transparent medium by the agent of illumination, and transmitted from the coloured surface to the eye. The account of perception given in the *De Anima*, while it is not put forward as a *justification* of the role of sense perception in knowledge,[38] does guarantee the veridical nature of sense perception, and hence provides strong support for Aristotle's reliance on this.

The process of discovering *archai*, through sense perception, is called *epagōgē*. It is clear when we consider the question of the relation between the *phainomena* and commensurate universals that simple *induction* (which is how *epagōgē* is usually translated) from sensible particulars would *not* suffice for the grasp of the latter. In the *Topics* we are told that 'it is by means of an induction of individuals in cases that are alike that we can claim to bring a universal into evidence'.[39] However, it is clear that inductive generalisations can only be complete over a finite domain and where we have access to every member of that domain.[40] Cases which fulfil these requirements occur rarely, and in physics for example they do not occur at all. Aristotle himself does not claim that commensurate universals are merely inductive generalisations based on a complete enumeration of all relevant instances for, he tells us in a later work, 'that which is commensurately universal and true in all cases one cannot perceive, since it is not "this" or "that" and it is not "now"'; also, 'induction proves not what the essential nature of a thing is but that it has or has not some attribute.'[41] A 'complete' induction, if that were ever possible, may lead to a proposition which is universally true, but it would not in itself lead to a commensurately universal proposition. That is to say, a complete induction could tell us that a thing is so, but not why it is so — it could not provide *aitia*. The distinction between that which is essential and that which is merely universal is a *sine qua non* of a theory of the demonstrative syllogism and hence it is crucial that Aristotle respect it. How then, we may ask, do we come by the *archai* which form the basis of demonstration?

In answering this question we must look more closely at that *epagōgē* involves. I have argued that it cannot be an enumerative abstraction. An alternative would be to say that it consists in penetrating beyond particulars to the universal. Such an interpretation receives some support, *prima facie*, from Aristotle's distinction between the orders of being and knowing. He distinguishes between that which is prior and better known in the 'order of being' and that which is prior and better known to man:

> 'Things are prior and better known in two ways; for it is not the same to be prior in nature and prior in relation to us, nor to be better known *simpliciter* and better known to us. I call prior and better known in relation to us what is nearer in perception, prior and better known *simpliciter* what is furthest away. What is most universal is furthest away, and the particulars are nearest; and these are opposite to each other.'[42]

However, support for this alternative rests on an ontological reading of this passage — a reading on which the order of nature is that of universals and the order of knowing is that of particulars. Such a reading cannot be correct, however, for in the *Physics* the same distinction is made but the contents of the two 'orders' are reversed:

> The natural course is to proceed from what is clearer and more knowable to us, to what is more knowable and clear by nature; for the two are not the same. Hence we must start thus with things which are by nature clearer and more knowable. The things which are in the first instance clear and plain to us are rather those which are compounded. It is only later, through an analysis of these, that we come to know elements and *archai*. This is why we should proceed from the universal to the particular. It is the whole which is more knowable by perception, and the universal is a sort of whole: it embraces many things as parts. Words stand in a somewhat similar relationship to accounts. A word like "circle" indicates a whole indiscriminately, whereas the definition of a circle divides it into particulars.'[43]

Basically, the kind of distinction being made here is not so much an ontological distinction as one between different 'forms' of knowledge.[44] This is indicated by the way in which we are said to know things 'better'. What is known at the starting point is known in an undifferentiated fashion. But it is not undifferentiated *simpliciter*; it is, rather, that the basis of differentiation does not adequately reflect the *kind* of thing which is differentiated. At this point we may recall our earlier discussion of universals. We noted there that some things which can be said of a subject are accidental whereas others are essential to that thing's being what it is. These latter predications give us the kind of thing something is and we have already seen that, for Aristotle, such kinds are natural. Natural kinds are self-differentiating, but this does not mean that our everyday speech automatically exhibits these kinds: this is part of the reason why demonstration is needed. Indeed, I think this is the whole point of the metaphor of the battle at the end of the second book of the *Posterior Analytics* (quoted above). What is being restored by the men making a stand is the original order (*archēn*). In his translation, Barnes substitutes *alkēn* (strength) for *archén*, claiming that the latter makes no sense.[45] But this substitution ruins an important metaphor, which is that *archē* means both 'principle' and 'original order'. The original order is the order of natural kinds, the principles of differentiation of which are *archai*. As Randall[46] has pointed out, *epagōgē* 'means seeing more than mere thats, not seeing less; it is not a process of denuding, but the discovery of additional meaning. It means seeing not only the particular instances, but seeing also the intelligible structure of the particulars that is implicit in the various thats'.

Finally, we must offer some clarification on the question of 'knowing better' or 'better knowledge' and this is best done by considering the concept of *nous*. Aristotle tells us that of the 'thinking states by which we grasp the truth' the most central and basic state is that of '*nous* which apprehends the primary premises'.[47] On the standard interpretation,[48] *nous* is an extra process over and above *epagōgē*. Now it is true that we could not know the primary premises without *nous*, but this in itself does not render *nous* a process. Just as in saying that it is by sight that we see what we are looking for, this does not mean that sight is a process by which we find what we are looking for, so in saying that it is by means of *nous* that we know the basic *archai*, this does not mean that *nous* is a process by which we come to know the basic *archai*.[49] Further, it would be quite unclear why *epagōgē* was needed at all if we possessed *nous* as a faculty. A more plausible interpretation is that *nous* is the *state* we are in when we have grasped the basic *archai*. What happens in *epagōgē* is that universals are grasped in a series which — to use Aristotle's example — goes from the universal 'animal', to the species 'animal', to unitary animals. These stages tend towards indivisible and unmiddled terms.[50] Since it is *nous* alone which can grasp first principles, *nous*, as Lescher[51] points out, can be 'truer' than *epistēmē* because it can grasp principles which are the *aitia* for other principles' being true. *Epagōgē* is the means by which we grasp universal principles, and these form a series the limiting case of which is first principles. These first principles are the basic *aitia* and, as such, are the definitions of natural kinds from which the essential properties of bodies can then be derived.

§3.1   Proof: Subject Genera

Up to now we have dealt with the concept of *archē* at a general level. Earlier, we noted that the special *archai* of the particular sciences determine the subject genera of those sciences. We shall now deal with this question in detail. In this section, we shall discuss the difference between the subject genera of physics and mathematics at a rudimentary level, and then (§3.2) we shall look more closely at Aristotle's concept of number (*arithmos*) and its bearing on his conception of mathematics. Finally, in §3.3 we shall discuss the difference between physics and mathematics in terms of their different *archai*.

In Aristotle's classification of the 'theoretical' (as opposed to the 'practical' or 'productive') sciences, *theology* or *metaphysics* (what Aristotle simply calls 'first philosophy') deals with those things which have a separate existence and are unchangeable, *mathematics* deals with those things which have no separate existence but are unchangeable, and *physics* deals with those things which have a separate existence but are changeable.[52] This classification of the theoretical sciences is not merely a heuristic device, nor is it an incidental feature of Aristotle's account; it is a substantial demarcation between domains of study which carries with it a demarcation between the kinds of *archai* appropriate to these domains. Furthermore, although the mode of proof common to all sciences is the demonstrative syllogism, the theory of the demonstrative syllogism involves a demarcation between the kinds of proof appropriate to the particular sciences.

In scientific demonstration, what are sought are commensurately universal relations. These are essential and universal relations between the major and minor terms of the syllogism, reflecting an essential and universal relation between a subject and its attributes. Phenomena must be explained in terms of essences, or *archai* construed as *aitia*. Explanations of this kind would constitute absolutely certain knowledge. If anything is in doubt in Aristotle's account of explanation it is whether the kind of explanations required by this account can be given: there can be no doubt that if explanations of the kind proposed could be given they would constitute absolutely certain knowledge.

In looking at this concept of explanation more closely it is important that we determine its relation to the question of what counts as a proof in the particular sciences. We shall deal with this question by examining the sense in which Aristotle conceives the theoretical sciences to be autonomous domains of study and, in particular, the sense in which mathematics and physics are autonomous domains of study.

On Aristotle's account, when the mathematician deals with 'surfaces and volumes, lines and points' he is dealing with attributes of physical bodies. The physicist also deals with these attributes, but whereas *he* treats them *qua* the attributes of bodies — as the limits of physical bodies — the mathematician treats them in their own right,

> 'nor does he consider the attributes indicated as the attributes of such bodies. That is why he separates them; for in thought they are separable from motion, and it makes no difference, nor does any falsity result, if they are separated.'[53]

The mathematician, when he is concerned with spheres, for example, is concerned with spheres as such, and not with spherical bodies or with sphericity *qua* an attribute of matter. The physicist is concerned with both matter and form, so he is concerned with sphericity *qua* attribute of a body: he is not concerned with sphericity as such but with

whatever is spherical. Hence in the *De Caelo* we are offered a
physical proof of the sphericity of the Earth:[54] such a proof
comes within the domain of study of the physicist only, not
within that of the mathematician. The attributes studied by
the mathematician, incidentally, are said to be separable (in
thought) from *motion* — and not from *matter*, as we might
initially expect — because strictly speaking the study of
nature is primarily the study of things subject to motion
(*kinēsis*) or change (*metabolē*). This is *ultimately* the study of
matter only because logical analysis reveals that a material
factor is presupposed in any motion or change, as we shall
see later.

§3.2   Mathematics and the Concept of Arithmos

Aristotle considers that the mathematician studies something which is *abstracted* from sensible things. In modern
terms, this is a very curious idea. We do not usually think of
numbers, for example, as being *abstracted* from sensible
objects. Nevertheless, such a conception is quite in keeping
with the Greek concept of number (*arithmos*). The account
we give of the objects of mathematics is clearly important if
we wish to determine what role, if any, mathematical
procedures can play in the proof of physical theorems. In
dealing with Greek mathematics one of the central problems
lies in the concept of *arithmos*, and in this section we shall
consider what exactly is involved in this concept.[55]

In giving the *arithmos* of something we are giving a
definite number of definite things. For Aristotle, just as
there is no triangle which is neither equilateral nor scalene, so
there is no decad which is not this or that ten definite
things.[56] A decad is always a decad *of* something, whether it
be dogs, sheep or whatever. Just as Plato speaks of numbers
which have 'visible and tangible body',[57] so too for Aristotle
'to be present in number is to be some number of an
object'.[58] The problems arise when we come to consider what
is involved in counting. We can 'count off' objects without
any difficulty, but this 'counting off' presupposes a
(logically) prior knowledge of numbers. It is the analysis of

this prior knowledge which presents the difficulties. Number is always the number *of* something. In the particular case, we count numbers of dogs and sheep, for example. But our prior knowledge of numbers is not of decads of dogs and sheep: exactly insofar as it is *prior* knowledge, what these numbers are numbers *of* cannot be the same as the sensible objects which we count off. In studying numbers *as such* — as opposed to counting dogs or sheep — numbers do not, and cannot, thereby cease to be numbers *of* something: what *does* happen is that they cease to be numbers *of sensible objects*. The objects of which these numbers are the numbers must fulfil two requirements: they must be *noetic* objects and not sensible objects, and they must exhibit all the essential characteristics of what is countable. These requirements are fulfilled by 'pure units' — or *monads* — which are accessible only to the understanding, indistinguishable from one another, and resistant to all partition.[59] These 'pure units' cannot be identified with 'rational numbers'. There are two reasons for this. First, 'one' is neither a 'pure unit' nor a number of any kind for the Greeks. We shall examine why this is the case below. Secondly, while it is true that there is a difference between a decad of sheep and a decad of pure units accesible only to thought, a decad of pure units is still a 'definite number of . . .'. For the Greeks, number is *always* related to that of which it is the number. There are no *arithmoi simpliciter* as there are rational numbers *simpliciter*.

This conception of *arithmos* is common to Plato and Aristotle. Where they differ is in their accounts of the 'pure units' or *monads*. The positing of such *monads* is required if an account of mathematics is to be given on the basis of the concept of *arithmos*, so the difference between the Platonic and Aristotelian doctrines cannot lie here. It lies, rather, in the 'mode of being' or *ontological status* of these *monads*.

Plato posits a Form of number — the *arithmos eidetikos*. In Aristotle's summary[60] of Plato's *arithmos eidetikos*, it is defined as a being with multiple relations to the other Forms but which nevertheless remains altogether indivisible. Plato appears to confer an independent being on those numbers of which we must have prior knowledge if we are to be able to

count. This is what Aristotle objects to.[61] For Aristotle, just
as the existence of colour is dependent on the existence of
what it is the colour of, so too is the existence of a number
dependent on the existence of what it is the number of, and
colours and numbers are *primarily* colours and numbers of
sensible things. For example, just as, when we make an
assertion about 'green trees', the existence of *green* is
dependent on the existence of the trees,[62] so too when we
make an assertion about 'three trees', *three* has no existence
outside of the trees of which there are said to be three: the
number of trees has no independent 'nature'.[63] Their being
'so many', just like their being 'green', is dependent on their
being trees.

The problem Aristotle must now deal with is this.
Arithmos has a 'natural significance' in that assemblages of
sensible objects have definite numbers. However, *arithmos*
also has a 'dependence' — it is dependent for its existence on
what it is the *arithmos* of. How is this dependence
compatible with 'pure numbers', which are purely *noetic?*
Aristotle answers this question by considering how we come
by pure numbers. The procedure involved here is exactly the
same as the procedure by which we come by any object of
epistēmē. We 'abstract' — if that is the right word (see §2) —
from sensible and changing things that which does not
change. In the present case, we abstract the numerical
aspects or dimensions from sensible objects. This renders
mathematical objects *noetic* and insensible, but it does not
confer on these objects an ontological status which is
independent of the sensible objects from which they are
abstracted. These mathematical objects are 'indifferent' to
the sensible things they are numbers of,[64] they are *neutral
monads*. Whereas for Plato any account of number must
begin with neutral monads — since these are ontologically
prior to numbers of sensible things — for Aristotle numbers
are transformed into neutral monads *only* when they become
an object of *epistēmē*. They have no ontological priority of
any kind.

On Aristotle's account, each of these neutral monads
represents 'a number by which we count'.[65] However, just as
the numbers of sensible objects are not 'one thing' so the

numbers of neutral monads are not 'one thing'. Numbers are
'one' only insofar as they extend over a 'whole'. We can only
ask what makes the number 'ten' *one* number by reference to
those things which are actually counted. In counting things,
we can only count things of the same kind, and it is insofar as
things are of the same kind that they are 'one'. *One* is the
'measure' by which we count: 'for number is a multitude
measured by a unit'.[66] Just as in mensuration we count off
units of measurement[67] so in counting we count in terms of
units which are 'neutral'. Further, we count '(one), two,
three . . .' and not '(one apple), two apples, three apples . . .'.[68]
but the former procedure is parasitic on the latter. The
pure mathematical unit — the *monas* — is something
abstracted from sensible objects; it is a *measure*. Because this
measure is a measure of homogeneity, and because homogeneity is a precondition of counting — in that we can only
count over homogeneous collections — the unit has a special
status. It is a precondition of counting: 'For each number is
"many" because each is [made up of] "ones" and because
each is measured by [its own] "one".'[69] It is insofar as things
'are one' that they are countable. In this sense the *one* takes
precedence over number, and becomes its *archē*.[70]

Arithmos signifies a 'numbered assemblage'. It is not
equivalent to number in the modern sense. It does not
designate a general magnitude. Aristotle's use of mathematical letters, in the *Physics* and in the *De Caelo* for
example, is not a *symbolic* use. This non-symbolic
conception of number is to be found also in the
'arithmetical' books of Euclid,[71] where the 'pure' units in
terms of which numbers are composed are taken as 'units of
measurement' such as can be represented most simply as
straight lines which are *directly measurable*. The nature of
each number — irrespective of whether it is 'linear', 'plane'
or 'solid' — is defined with respect to the measuring
character of its factors. This renders 'indivisibility' a
property of numbers not *qua* self-subsisting things — which
would ultimately render fractions inaccessible to analysis, at
least on Plato's account — but *qua* units of measurement.
Because units of measurement can be changed, fractions
no longer present a problem: we simply transform fractional

parts of an original unit into 'whole' numbers consisting of new units of measurement. This enables the solutions of calculations to be presented in terms of neutral monads. *But*, such solutions must always be determinate, since *arithmos* always designates a determinate magnitude. This precludes the setting up of *general equations* with *general* solutions. Similarly, we find negative and irrational numbers being treated as unacceptable solutions to calculations in the work of mathematicians up to and including Diophantus. These are 'impossible' numbers and their appearance as the solution to a problem is taken to indicate that the problem has been posed incorrectly.[72]

We can see, then, that the concept of *arithmos* produces severe computational problems which seriously restrict calculation. But more importantly, for present purposes, the domain of investigation of mathematics is construed in such a way that number, and magnitude, are abstractions from sensible objects. Mathematics is possible only because of the homogeneity of collections of such objects. This means that mathematics cannot be used to articulate this homogeneity. It is only in giving the *archai* of *phusis* that an account of this homogeneity can be presented, and this is the domain of physics. We are now in a position to return to the question of the relation between physics and mathematics in terms of their *archai*.

§3.3 Proof: *Archai*

Up to now, we have noted that, on the Aristotelian account, mathematics and physics have distinct subject genera: the one deals with attributes in their own right and the other with attributes *qua* attributes of something. The matter does not end here however. Distinct subject genera require distinct *archai*:

> 'It seems that perceptible things require perceptible principles, eternal things eternal principles, corruptible things corruptible principles; and, in general, every subject matter principles homogeneous with itself.'[73]

This requirement is *a consequence of the theory of the demonstrative syllogism*:

'Since it is just those attributes within every genus [kind] which are essential and possessed by their respective subjects as such that are necessary, it is clear that both the conclusions and the premisses of demonstrations which produce scientific knowledge are essential . . . It follows that in demonstrating we cannot pass from one genus to another.'[74]

Since each science is defined in terms of its genus, and since in the demonstrative syllogism what are demonstrated are the essential properties of the subject genus, the principles of one science cannot be used in the demonstrations of another. In the demonstrative syllogism we are concerned with attributes which inhere essentially in a subject. Where subject genera are distinct, distinct *archai* are needed. For example, geometry only deals with lines *qua* lines; in virtue of its subject genus, it can tell us nothing about lines *qua* boundaries of bodies, nor can it tell us, for example, that the line is the contrary of a circle.[75] Because form and matter are only separable in thought and not in reality,[76] it is not the case that the mathematician — who deals with forms — somehow covers only half the subject genus covered in physics — form and matter. Were the two separable in reality this would be the case; however, since a form *qua* essential attribute of matter and a form *simpliciter* are generically distinct, form and matter cannot be separable in reality.[77] The conclusion is that mathematical proofs are inappropriate in physics.[78]

There are two situations which, *prima facie*, seem to tell against this account inasmuch as they seem to indicate that Aristotle followed a different procedure in practice than the one which he puts forward at the programmatic level. They are, however, both quite compatible with his general account. The first concerns those cases where the theorems of one science can be demonstrated by the principles of another. These are the cases where the theorems to be proved are part of a subordinate science, and where these theorems are proved by means of the respective superior science.[79] For example, Aristotle considers that optics is a science which is subordinate to geometry. The reason for this is that although the subject-genera of optics and geometry are distinct, the 'reasoned fact' — the account in terms of basic *aitia* —

'concerns the superior science, to which the attributes essentially belong'.[80] In general, the subordinate sciences are subordinate to physics in that the subject of any of them is a species of the subject of physics: what differentiates them from physics is the fact that the attributes they investigate are not, as such, properties of the subject, whereas the attributes examined in physics are properties of the subject. They are subordinate to mathematics in that the attributes which they investigate are not properties of the subjects of those subordinate sciences but are the subject or the properties belonging to mathematics. The existence of the subordinate sciences does not go against the thesis that attributes are only demonstrative from 'appropriate' basic truths since these sciences have the 'requisite identity of character'.[81] Indeed, 'optics investigates mathematical lines, but *qua* physical, not *qua* mathematical'.[82] Further, the *archai* involved in the subordinate sciences cannot be either wholly mathematical or wholly physical, otherwise we would have an invalid (because non-homogeneous) proof. Neither can one of the premisses be physical and the other mathematical for in this case there could not be a middle term: this is clear when we consider that mathematical premisses contain two mathematical terms and physical premisses two physical terms. The basic kind of *archē* which is appropriate in the subordinate sciences is one which a physical subject ·and a mathematical predicate. Any subordinate 'demonstration' is only a demonstration insofar as its *archai* are parasitic on the *archai* of physics and mathematics: any necessity derives from these, not from the subordinate science itself.

Secondly, although the *archai* must be homogeneous with the subject-genus in a demonstration, and hence peculiar to their genus, this does not mean that the *archai* of different genera are incommensurable (incommensurable in the sense that it is meaningless to speak of incompatibility). Mathematical principles cannot be used in a physical proof, but if a consequence of a physical theorem is incompatible with a mathematical theorem then one of the theorems is false. This means that a situation can arise where although a physical theorem cannot be proved mathematically, it can be disproved mathematically. For example, in the *De Caelo*,

Democritus' atomism is rejected on the following grounds (amongst others):

> 'A view which asserts atomic bodies must needs come into conflict with mathematical sciences, in addition to invalidating many common opinions and apparent data of sense perception.'[83]

Such a situation is not as paradoxical as it seems if we distinguish two notions of proof (and hence of disproof). We can prove that a thing is so and we can prove why it is so. In physical demonstration we prove why things are so. The mathematical disproof of a physical theorem can only tell us *that* a thing is not so (in the way that sense perception can tell us this) and not *why* it is not so.

§4   Explanatory Structure

In summary, we may describe the concept of explanation which Aristotle works with as follows. The theoretical sciences, in which explanations are given, are defined or differentiated in terms of their subject genera. To explain something is to demonstrate sensible phenomena (including 'occurences') from their *aitia*. In explaining a phenomenon we must give an account of it in terms of essences, or *archai* construed as *aitia*. Explanation is demonstration from first principles, and it takes the form of universal propositions which are necessarily true. These propositions — statements of commensurate universals — can only be demonstrated by *archai* which are homogeneous with the subject genus, since we could not derive the essential attributes of a subject from non-homogeneous principles. Since the subject genera of mathematics and physics are distinct, the *archai* of one cannot be used in a demonstration of the other.

The principles of syllogistic reasoning — expounded in the *Prior Analytics* — and the principle of homogeneity — expounded in the *Posterior Analytics* — constitute the conditions under which a *proof* can be valid or invalid. This use of the concepts 'proof' and 'validity' may strike the reader as eccentric since whether *archai* are homogeneous with their subject genus is something to be decided in *epagōgē*, and hence ultimately by sense perception, whereas the concepts of 'proof' and 'validity' are normally reserved for describing the formal structure of a system only. But in fact the usage does not depart from this requirement. Heterogeneous principles are not like untrue axioms or theorems; rather, they are *inappropriate* principles. While homogeneity may be a requirement of the formal structure of a system, this does not entail that *the criteria for* homogeneity are part of that formal structure. The principle of homogeneity — which is a common *archē* — is a prerequisite of proof, but the principles by which we decide homogeneity are the *archai* peculiar to the particular science we are considering.

Let us now return to the question of evidence. The phenomena with which the physicist deals are those of sense perception, either in the strong sense — phenomena which can actually be perceived, such as motion — or in the weak sense — phenomena which can be deduced from what we see, and which have their fundamental properties (weight/levity, extension, determinate position, duration in time etc.) in common with the objects of perception: such as the four elements. Sense experience is the arbiter of the truth of physical theories in the sense that the consequences of these theories must not contradict what we perceive under normal conditions. 'Under normal conditions' means that neither the perceiver, not the medium which transmits the light, should be affected in any way. For example, the former should not be ill, drunk or asleep; the latter should be pure transparent medium in which there is no interfering body composed of other elements such as water or earth (this qualification has an important bearing on the use of lenses, as we shall see in the next chapter).

Within this explanatory structure, then, the domain of evidence in the theoretical sciences, and in physics in particular, is restricted to sense experience. This is not surprising when we consider what is being explained. The objects of explanation are the phenomena as these are perceived. In giving the *aitia* of these phenomena we are simply providing a rationale (*logos*) for their differentiation. Because this *logos* actually inheres in the things themselves (insofar as they are known and named) the differentiation is not a conventional one; rather, it is a recognition and articulation of the self-differentiating kinds of phenomena. On this basis we can then deduce the essential properties of the things belonging to each kind. The crucial issue here is how we recognise natural kinds without this simply turning into abstractionism. Aristotle's account of *epagōgē* purports to tell us how we could possibly have knowledge of *essential* and universal relations, but it does not even purport to tell us how we are actually to come by these relations. As a description of how we come by *universal* relations it works in its own terms, but even then these universal relations are only generalisations which, problems of necessity aside, could equally well be accidental as they could be nomological.

Now suppose that we argue that although *epagōgē* can only be guaranteed to lead us to universal statements, these statements may, *as a matter of fact*, also be *essential*. The problem here is that we could not establish whether they are essential or not on the basis of *epagōgē*. This means that we could never know whether any universal statement was essential, and *in practice* we would have to work with generalisations which, by Aristotle's own criteria, would be inadequate.

This brings us to the central issue: are explanations of the kinds required by Aristotle possible? In the next sections we shall examine Aristotle's physics and cosmology. The problem we shall deal with is this. In order to give explanations of the required kind in physics, we must know the natural order of *archai* or essences. Such knowledge is clearly not the same as that derived by abstraction from individual cases: abstraction can provide us with knowledge of accidents but not necessary attributes as such. If we find that the search for essences can in fact only take the form of abstraction from individual cases then Aristotle's account of what counts as an explanation in physics is an unrealisable ideal. I shall argue that this is in fact the case and that his project is abstractionist, which means that it is essentially and fundamentally misguided, since it means that, by his own criteria, truly scientific knowledge cannot be attained, at least in physics and cosmology.

§5   The Pre-Aristotelian Problem of Motion

Book III of Aristotle's *Physics* opens:

> 'Nature has been defined as a "principle of motion and change", and it is the subject of our enquiry. We must therefore see that we understand the meaning of "motion"; for if it were unknown, the meaning of "nature" too would be unknown.'

The problems that Aristotle poses with respect to motion are new ones. In this section we shall examine what problems of motion had been posed before Aristotle, and we shall try to determine in what way the problems shift between Parmenides and Aristotle. We must do this because before examining how Aristotle carries out his project in physics it is important that we determine how this project is conceived. The first question we must look at is why Aristotle thinks an account of motion *can* be given.

Parmenides had denied that movement, or, more generally, change occurs; *a fortiori*, this precludes movement or change being an object of study in its own right. He did not deny the phenomenal reality of change, but argued that since a thing either is or it is not, then change cannot have any substantial reality. The relegation of change to the realm of 'the beliefs of men' meant that in the study of physics we are no longer in the realm of 'trustworthy discourse and thought concerning truth'.[84] 'Thought concerning truth' must be based on eternal and indestructible principles, and it must concern itself with the eternal and indestructible.

Now although Plato accepted this maxim,[85] he did not construct his principles in such a way as to eliminate change. He admits that those things which undergo change cannot be an object of knowledge, but it is nevertheless suggested that the relation in which these things stand to the Forms may serve for the description and characterisation (by contrast) of change.[86]

There are two questions which need examination here. First, why cannot a change or movement be an object of knowledge for Plato? Second, in what sense can the change or movement of things be an object of study nevertheless?

With regard to the first question, we may note that the Forms themselves are changeless: they are no more dependent on particulars for their nature than for their existence. However, although the Forms themselves do not change, and although change itself cannot be made subject to the Forms, in later works Plato does introduce a Form of change itself. In the *Sophist*, movement (*kinēsis*) is made one of the Forms, and indeed one of the 'Greatest Kinds'.[87]

Particulars, on the other hand, are transient and in a state of continual change. This means that they cannot be the object of genuine knowledge, since genuine knowledge is knowledge of the essential natures of things — the Forms — and that which changes cannot have an *essential* nature inasmuch as it changes. The sensible world, being sensible, is in a state of 'becoming' or change, and in the *Timaeus* we are told that whatever 'becomes' has a cause, that is, is the product of an agent.[88] Agents such as craftsmen always work with a model or archetype, and since the agent responsible for the Cosmos is the best of all causes, the model of which the sensible world is a 'copy' is eternal and unchanging.[89] The model can be the object of knowledge, but the material Cosmos, which is changing, can only be the object of *doxa*: 'As being is to becoming, so truth is to belief'.[90] In any account of the sensible Cosmos — or, more specifically, the sensible realm of nature (*phusis*) — the best we can aim for is to 'adduce probabilities as likely as any others'.[91] In saying that our account can only be 'probable' Plato is not using the word in the sense of 'probably true', since *truth* is not applicable to the Cosmos, only to its model. Strictly speaking, the word should also not be understood pejoratively, in the sense of 'merely probable', since this suggests a probationary status *vis-à-vis* truth as a standard.

The ontological distinction between the Forms and *phusis* (*qua* an object of sense perception) is backed up, then, by an epistemic distinction: one is properly the object of knowledge, the other properly the object of belief *per se*. Mathematics — which has an important role to play in Plato's account of *phusis* — poses something of a problem in this respect. Epistemically, it seems to be an intermediary between the sensible world and the Forms. In the *Republic* the enduring status of theoretical geometry is contrasted with the limitations of its practical manipulations, and geometry is conceived as that which lifts the soul towards truth and reality — which lie beyond geometry.[92] Pedagogically, Plato insists that the study of mathematics prepares the student for dialectic, and hence for apprehension of the forms, insofar as it transcends sense perception. But the *epistemically* intermediary status of mathematics does not entail that it has an *ontologically* intermediary status. Aristotle and the neo-Platonists interpreted Plato as arguing in this way but — at least in the case of the *Timaeus*, which is our main concern — Plato does not posit, nor is he strictly committed to positing, number and figure as an intermediary realm of existence between sensible things and the Forms.

Since those things which undergo motion or change have such a peculiar epistemic status, we may well ask what 'probable' account can be given of them. In the *Timaeus* a distinction is made between *genesis* ('becoming') itself and the particular objects which it 'creates'.[93] The former is an essential and permanent characteristic of nature, whereas the latter are not. Fire, which is taken as an example of the latter, 'we must not call "this" or "that", but rather say that it is of "such a nature" '.[94] *Genesis* itself, on the other hand, is conferred with the highest status: 'My verdict is that being and space and becoming, these three, existed in their three ways before the heaven'.[95]

No explicit relation between *genesis* — becoming — and *kinēsis* — movement or change — is postulated in the *Timaeus*. *Genesis* is not considered as a form of something, but nor is it a phase or the upshot of *kinēsis*. However, when the Forms have been 'imposed' on the original chaos, when the receptacle is replete with its elements, a motion arises in both the formations of these elements and in the receptable itself.[96] Plato traces this motion to the lack of balance between the 'powers' inherent in the new formation. The contents of the receptacle impart a swaying motion to it, and the receptacle in turn moves the contents and produces a separation of the four elements, with the result that each of them tends to take a place of its own. This physical movement presupposes the existence of physical entities and hence must be preceded by *genesis*: *genesis* is prior to *kinēsis* in the sense that if no physical objects had come into existence there would be nothing to move.

This physical account of *kinēsis* is supplemented by a mathematical account. In the mathematical account, basic triangles produce or 'beget' solid figures. This is not conceived as a physical process: the mathematical lineage of the elements is meant to account for their behaviour insofar as their interactions are determined by their geometrical nature. Now it is *genesis* that differentiates the elements geometrically, and their interactions are treated initially in terms of their specific natures and shapes. Nevertheless, it is clear that in order for the elements to interact they must be in motion: *kinēsis* is prior to *genesis* in the sense that if movement were not kept up in the cosmos the elements would have separated and nothing new would have come into being.[97] Plato's heuristic treatment of *genesis* before *kinēsis* is justified on the grounds that 'unless a person comes to an understanding about the nature and conditions of rest and movement [an understanding presumably provided by the account of *genesis*], he will meet with many difficulties in the discussion [of rest and motion] which follows.'[98]

In introducing *kinēsis*, Plato merely explains, in terms of *kinēsis*, those interactions between elements, and the resulting transformations, which he had previously established in terms of *genesis*. On the *genesis* account, it was established that changes from one element to another are possible owing to their specific nature and form. But this would not be 'casually' sufficient for the change actually to occur, and hence we need a 'source of movement' which explains what 'causes' the elements to collide. What this source of movement is becomes clear when we consider that all cosmic movements are produced by the revolution of the outermost heaven; that is, the world-soul, since rotation is a manifestation of the world-soul.[99]

Plato's account of cosmogony in the *Timaeus* — introducing, as it does, both physical and cosmological questions — is particularly instructive in respect of his treatment of *kinēsis*. This treatment is twofold: we are given a mathematical (in this case geometrical) account and a more properly 'causal' account — in terms of the soul (*nous*).[100] What is noticeable is the absence of the Forms. Although the true and ultimate *phusis* of things is the Forms, *phusis* also denotes the realm of movement (ie., the realm of things which move, as opposed to movement itself, which is a Form), which contrasts with and complements the realm of rest (ie., the realm of Forms). Forms figure in the account of *genesis* and they are a necessary condition of the existence of the physical entities which are copies of them. But the twofold account of *kinēsis* does not, and need not, make reference to the Forms. The Forms do not cause *kinēsis*. The causal factor is *nous*, and the conditions under which this causality is possible are specified in geometrical terms.

Aristotle's account of *phusis*, and of *kinēsis* in particular, differs from Plato's in three important respects. First, we have noted that the accounts of natural processes which are put forward by Plato in the *Timaeus* are considered epistemically distinct from his more properly 'metaphysical' enterprises: an index of this is the absence of the Forms in these accounts. For Aristotle, on the contrary, the study of natural processes is not epistemically distinct from any other kind of study, either with respect to the certainty of its results or with respect to its legitimacy as an independent domain of study: natural processes, and especially motion, can be known and indeed *must* be known if any account of nature is to be possible. Secondly, Plato's world-soul is eliminated from Aristotle's account of *kinēsis*, although the soul is retained in his account of the movement of living beings. Thirdly, Aristotle replies to the Parmenidean denial of change explicitly, and he argues for the necessity of a substratum which underlies all change, thus making change a process which is subject to rational explanation.

§6 Aristotle's Treatment of Motion

For Aristotle, the physicist — the 'student of *phusis*' — deals with those things which are subject to change. These things have in themselves the source of their changing or remaining unchanged.[101] We have noted above that on the Aristotelian account it is a *sine qua non* of any account of nature that we be able to give an account of motion. The establishment of physics as a fully independent domain of study thus requires that both the meaning and existence of motion be established. If the existence of motion, and change in general, is to be established conclusively, then Parmenides' argument in the *Way of Truth* must be refuted. Aristotle's general project is to reconcile 'being' with the idea of change or motion. He does this by distinguishing change from what is changed. Any change requires a substratum which persists throughout that change. Change itself must not be confused with what is changed or with the result of change: it is because of a confusion of the first kind that Parmenides' argument leads to such an absurd consequence.[102] For example, on Aristotle's account, in saying that the air changes from hot to cold we do not mean that the air itself changes; rather, we mean that one of its qualities changes. The necessity of a persisting substratum follows from the Aristotelian doctrine of categories, one of the premises of which is that accidents exist and are knowable only if substance exists; the doctrine requires that accidents be attached to substance.

The relation between the persisting subject and its qualities can most usefully be examined by looking at the two central theses on which the theory of *kinēsis* hinges. These two theses are (i) the doctrine of matter and form, and (ii) the doctrine of actuality and potentiality.

The form of an object is the set of its attributes. Form cannot exist independently, it must be the form *of* something; moreover, changes of form require a persisting subject. This subject is matter. The actual being of a subject is due to its form: its form is the principle of its unity. The potential being of a subject is also due to its form, in the sense that its present form will determine what kind of thing it can be or become. It is only by reference to its future form that a thing can be said to have a potential existence. The fact that matter has an actual form and many potential forms explains why both living *and* non-living beings have both actual and potential existence. This is very important, for it enables Aristotle to define nature (*phusis*) and natural objects as having the *archē* or *aition* of motion in themselves, and hence he can eliminate Plato's world-soul while at the same time conferring on nature all the properties of the latter:

> 'Of things that exist, some are due to nature and some are due to other *aitia*. The animals and their parts, and the plants and the simple bodies (earth, fire, air, water) are of the former sort for we say that these and the like are due to nature. All the things mentioned present a feature in which they differ from things which are not constituted naturally: each of them has *within itself* a principle of motion and of stationariness (in respect of place, growth or decay, or by way of alteration). On the other hand, a bed or a coat or anything of that sort, considered as satisfying such a description — i.e. insofar as it is the product of an art — has no innate tendency to change. But insofar as, and only insofar as, it happens to be composed of stone or earth or of a mixture of the two, it *does* have such a tendency. This suggests that *nature is a source or cause of being moved and of being at rest in that to which it belongs primarily*, in virtue of itself and not in virtue of any concomitant attribute.'[103]

Solmsen[104] has noted the analogies between this doctrine and the general doctrine of *archai*. It is not some specific contraries which are principles, as Anaximander had asserted, but simply contraries as such, defined as form and privation. Analogously, it is not any specific physical substance which is the *archē* of motion but *phusis* as such. The doctrine of nature as the *aition* and *archē* of motion is one of most central tenets of Aristotle's physics.[105]

Before examining the doctrine of actuality and potentiality it is important that we distinguish what kind of change or motion we are dealing with. Motion or change in its most general sense is termed *metabolē* by Aristotle. *Metabolē* in respect of substance is either generation or corruption: a substance *X* becomes a substance non-*X* or vice-versa. *Kinēsis* is that *metabolē* which affects the state of an existing substance, and it comprises change or motion in respect of quality, quantity or place. In the case of *kinēsis*, a substance *X* acquires a property *p* which it did not have previously. The crucial difference between the two kinds of *metabolē* lies in the nature of the *termini* between which it takes place: in *kinēsis* the termini are contraries, in generation and corruption they are contradictories. It is *kinēsis* that we shall be concerned with.

Kinēsis is defined as 'the *entelecheia* of that which is

potentially [*dunamei*], as such'.[106] To translate the term *entelecheia* here as 'actuality' would seem to suggest that *kinēsis* is a product or resultant of a process, whereas Aristotle is clearly speaking of a process of some kind, and not its product. If, on the other hand, we take *entelecheia* to mean 'actualisation', and construe the definition as meaning that *kinēsis* is the process by which something potential is made actual, then the meaning of the 'as such' (*ē toiouton*) becomes obscure, for not every actualisation is a *kinēsis*: the actualisation of a man's disposition to be wise — that is, the manifestation of this wisdom — is not a *kinēsis* for example.[107] So if not every actualisation of a potentiality is a *kinēsis*, then *kinēsis* cannot be defined as the actualisation of a potentiality, as such — at least if the 'as such' applies to the actualisation. The problems are compounded when we are told that '*kinēsis* seems to be a kind of *energeia*, but *atelēs* [incomplete or imperfect]',[108] for *energeia* usually means 'actuality', which simply brings us back to the first option.

It is clear that there are two distinctions which require examination if we are to begin to understand the definition. They are the actuality/potentiality (*energeia/dunamis*) distinction and the *kinēsis/energeia* distinction.

In the *De Anima*, a distinction is made between two senses of 'actual' and two senses of 'potential':

> 'But we must now distinguish not only *between* what is potential and what is actual but also different senses in which things can be said to be potential or actual; up to now we have been speaking as if each of these phrases had only one sense. We can speak of something as "a knower" either (a) as when we say that man is a knower, meaning that man falls within the class of beings that know or have knowledge, or (b) as when we are speaking of a man who possesses a knowledge of grammar; each of these is so called as having in him a certain potentiality, but there is a difference between their respective potentialities, the one (a) being a potential knower, because his kind or matter is such and such, the other (b), because he can in the absence of any external counteracting cause realise his knowledge in actual knowing at will. This implies a third meaning of "a knower" (c), one who is already realising his knowledge — he is a knower in actuality and in the most proper sense is knowing, e.g. this *A*'.[109]

Following Kosman,[110] these distinctions can be clarified by considering the case of speaking Greek. A normal Athenian baby (*A*) is capable of being able to speak Greek. A normal

Athenian adult who is silent (*B*) is able to speak Greek. A normal Athenian adult who is speaking his own language (*C*) speaks Greek. *A* and *B* potentially speak Greek but in different senses. *B* and *C* actually speak Greek but in different senses. We can say that *A* potentially$_1$ speaks Greek, that *B* potentially$_2$ and actually$_1$ speaks Greek, and that *C* actually$_2$ speaks Greek. Hence, *A* has a potentiality which is realised in *B* in the form of an actuality which is itself a potentiality (and itself realised as an actuality in *C*).

Let us now return to *kinēsis*. The definition of *kinēsis*, we are told, can be elucidated by examples, and the main example that is provided is that of building:

> 'Take for instance the buildable as buildable. The actuality of the buildable is the process of building. For the actuality of the buildable must be either this or the house. But when there is a house, the buildable is no longer buildable. On the other hand, it *is* the buildable which is *being* built. The process then of being built must be the kind of actuality required. But building is a kind of *kinēsis*, and the same account will apply to the other kinds also.'[111]

That is, bricks and stones manifest their potentiality as buildable (i.e. as building materials) in the building of the house, not in the house itself, for once the house has been built the buildable has been actualised in such a way that it is no longer buildable. Thus we can say that the actuality$_1$ of the buildable is the process of building; the actuality$_2$ of the process of building is the house. Now this case must be distinguished from that in which we build purely for the pleasure of building. In the first case, the ultimate end of the activity is something which is other than itself; in the second case, the ultimate end of the activity is something which is not other than itself.

This distinction is an important one since the first case is an example of a *kinēsis*, whereas the second is an example of an *energeia*. This is the point of the following passage from the *Metaphysics*:

> 'At the same time we are seeing and have seen, are understanding and have understood, are thinking and have thought: while it is not true that at the same time we are learning and have learnt, or being cured and have been cured.'[112]

Seeing, understanding and thinking are *energeiai*; learning and being cured are *kinēseis*. Our first example — that of speaking Greek — is an example of an *energeia*. We have noted that both *energeia* and *kinēsis* can be elaborated in terms of the two kinds of actuality and potentiality. What is distinctive about *kinēsis* is that it is the manifestation of a potentiality *qua* potentiality. When we are told that *kinēsis* 'is a kind of actuality [*energeia*], but incomplete [*atelēs*], the reason for this view being that the potentiality whose actuality it is is incomplete',[113] this cannot mean that *kinēsis* is to be distinguished from other forms of *energeia* — or more strictly speaking from *energeia* proper — in virtue of its being a special kind of incomplete potentiality, the actuality of which is incomplete and is *kinēsis*, for all potentiality is incomplete by definition. Rather, as Kosman notes, the point being made here is that *kinēsis* 'is the constitutive actuality of an entity which is by definition incomplete, and since the constitutive actuality of an entity is simply that entity in its full manifestation, *kinēsis* itself is an incomplete actuality'.[114] *Kinēsis* is incomplete because it is the actuality$_1$ of a potentiality$_1$: but in being an actuality$_1$ it is also a potentiality$_2$ and the actuality$_2$ of this potentiality$_2$ is something which is not only other than *kinēsis*, it is the annihilation of *kinēsis*.

Now since it is not an essence, *kinēsis* itself cannot strive towards an end — only that which undergoes *kinēsis* can do that. This means that one change cannot directly generate another: a *kinēsis* cannot be at either of the termini of a *kinēsis*. *Kinēsis* is not independent, it is something that occurs in a subject: it is a 'process' which a subject undergoes. There are two points to be noted here. First, *kinēsis* is generically distinct from rest. Rest is a state of body. The rest of a body in its natural place is a natural rest. An external agency is required if the body is to move from its natural place. The rest of a body which is not in its natural place is a state of privation. The body has a tendency to move to its natural place: that is, it has a tendency to undergo a process which will change its state. But again some agency is required if this process is to be initiated and maintained for, as we have just seen, *kinēsis qua potentiality* is directed towards its own annihilation. Now states do not require explanation.[115] Processes, on the other hand, do require explanation, and since processes are always the product of an agency this explanation must be 'causal'.

Secondly, in *kinēsis* potential qualities are 'actualised'. This presupposes a *terminus a quo* and a *terminus ad quem*. Since *kinēsis* is the process by which an object changes its properties, it is characterised in terms of the two contrary states between which the process occurs: the state of not having a particular property and the state of having this property. The specification of termini is important not only for the purposes of classification, it is an integral part of the concept of *kinēsis* (and *metabolē* in general).

Local motion is *kinēsis* in respect of place. We must distinguish between the *common place* of a thing — that place which it shares with other things — and the *proper place* of a thing — what immediately contains it. This is the first, and preliminary, definition of proper place. The second definition is that place is 'the boundary of the containing body at which it is in contact with the contained body', where the 'contained body' is 'what can be moved by way of locomotion'.[116] Place is distinct from shape, since shape is the boundary of the contained body, whereas place is the boundary of the containing body. In more modern terms, we would say that place and shape have the same extension but different intensions. Place is distinct from matter since matter is neither separable from the thing nor does it contain it. It is also distinct from 'what is in between bodies', their separation, since the distance between the extremities of bodies does not exist by itself; it is an accident of the bodies which are contained in the containing body. Finally, we come to the third definition of (proper) place:

> 'Just as, in fact, the vessel is transportable place, so place is a non-portable vessel. So when what is within a thing which is moved, is moved and changes its place, as a boat on a river, what contains plays the part of a vessel rather than that of place. Place on the other hand is what is motionless: so it is rather the whole river that is place, because as a whole it is motionless. Hence we conclude that *the innermost motionless boundary of what contains is place.*'[117]

Owen has noted that the concept of place as surroundings 'is normal in Greek philosophy, as the arguments of Zeno and Gorgias show (and in ordinary conversation, which has small use for plotting objects by Cartesian coordinates, it is still so). Aristotle took it over as an *endoxon* and made a more sophisticated version of it in the fourth book of the *Physics*'.[118] This is particularly instructive. It is true that Aristotle's account of place is dialectical in that he formulates the concept on the basis of critical reflections on previous theories,[119] and this indeed is Owen's point. However, the views he criticises, and his own progressive definitions of place have a common feature: they are all constructed on the basis of everyday experience. It is axiomatic to Aristotle's criterion of explanation that scientific knowledge be produced on the basis of sense perception.

An important feature of place is that it is absolute; a thing is not in a particular place with respect to some other thing, it is in a particular place *per se*. Aristotle's concept of absolute place is a necessary condition of his concept of absolute direction. The concept of absolute direction entails that of absolute place. In his finite, spherical, hierarchially-ordered Cosmos six absolute directions are distinguished: up and down, right and left, and forward and backward. The latter two pairs are absolute since they are defined in terms of the movement of the stars.[120] A definition of this kind both agrees with sense perception and is quite plausible. It is plausible because it does not matter where the observer is placed since motion — a process — is always distinct from rest — a state — so although things might *appear* differently if one were standing on one of the stars, this star would still *really* move and the earth would still really be stationary. One can have as many optical frames of reference as one likes, but this does not mean that there is more than one physical frame of reference. Aristotle cannot deny *optical* relativity — otherwise he would be committed to asserting that fire travels downwards when one is standing on one's head — but on his conception of the distinction between motion and rest he cannot allow *physical* relativity.

Now not only is place 'something, but it also exerts a certain influence'. This influence can be seen most clearly if we consider the places 'up' and 'down'. Up and down are not merely relative, for *in nature* up and down are distinct places, since 'it is not every chance direction which is "up", but where fire and light are carried'. This idea is closely connected with the doctrine of elements, which we must now examine.

Aristotle introduces four elements initially. Earth is associated with downward rectilinear motion and fire with upward rectilinear motion. Air and water are 'intermediate' between these. Water is 'potentially light' and air is 'actually light'.[121] We can distinguish two situations for a light body:

> 'The activity of lightness consists in the light thing being in a certain situation, namely high up: when it is in a contrary situation, it is being prevented from rising . . . [Light and heavy things] have a natural tendency respectively towards a certain position: and this constitutes the essence of lightness and heaviness, the former being determined by an upward, the latter by a downward tendency.'[122]

The four elements of the sublunar region each have a natural place. An element in its natural place has no tendency to move. The same holds for any body, since bodies are composed of the elements: the behaviour of a compound body is determined by that element which predominates in its composition. Natural motion is the motion of a body to its natural place. This is a rectilinear motion, and the body will come to rest when it has reached its natural place. If a body is made to move, by an external force, in any other direction than this, it is said to be in violent (or 'constrained') motion. Whereas natural motion proceeds from an internal *arche*, violent motion presupposes the action of an external impulsion.[123]

The idea of natural motion presupposes the idea of the Cosmos as a finite plenum, since there could be no natural motion in whatever is infinite or void, 'for insofar as it is a void, up differs no whit from down'.[124] There would simply be no reason for a body to undergo natural motion in whatever was void or infinite. Further, since on Aristotle's theory the speed of a body's motion varies inversely with the density of the medium, a void would be traversed instantaneously. Such motion Aristotle considers to be impossible. He concludes that there could be no natural motion in a void.

Violent motion in a void would be unmoved motion — which is again impossible. For Aristotle there can be no motion in the absence of a mover; thus, in dealing with the continued motion of a projectile he follows Plato in explaining such motion in terms of the reaction of the surrounding medium.[125] In justifying the part played by the medium in violent motion, Aristotle invokes the quality of the two intermediate elements — air and water — of possessing gravity and levity at the same time, which allows them to transmit motion in any direction. The medium is disturbed by the mover and carries the projectile along. The medium itself is not set in motion, but it is endowed with the faculty of imparting motion — a motive power which overcomes the tendency of the projectile to return to its natural place.

The most striking feature of this account is its sheer inconsistency. The medium resists natural motion but is responsible for the continuation of violent motion. This is not the only problem. As Galileo[126] was quick to point out, a ball falling through a hole drilled in the earth and passing through its centre would undergo natural motion to the centre but, because of the speed it would have acquired at this point, it would pass the centre. But in doing so it would be effecting violent motion, so 'a moveable body may be moved with contrary motions by the same internal principle'. That the one thing can be responsible for both natural and violent motions is a serious internal inconsistency in Aristotelian mechanics, and it reflects the fact that two central concepts — natural and violent motion — cannot be articulated properly on that theory.

From our discussion up to now, it will be clear that the doctrine of elements is of absolutely central importance in the Aristotelian account of motion. The existence of the elements is 'deduced' in Book III of the *De Caelo*, and in Book IV the doctrine of elements is fully established. Each element is associated with a specific movement and the elements are studied almost exclusively with respect to their movements. This association is approached via the issue of heavy and light qualities. These qualities reside in the elements themselves and are particularly important on account of their relation to the natural movements:

> 'By absolutely light I mean [a body] which of its own nature always moves upwards, and by absolutely heavy one which of its own nature always moves downward, if no obstacle is in the way.'[127]

Aristotle tells us that there is general agreement on the existence of absolutely heavy bodies. Evidently, this agreement does not extend to absolutely light bodies, so we are provided with the following argument:

> 'We see with our eyes, as we said before, that earthy things sink to the bottom of all things and move toward the centre. But the centre is a fixed point. If therefore there is some body which rises to the surface of all things — and we observe fire to move upward even in air itself, while the air remains at rest — clearly this body is moving towards the extremity. It cannot then have any weight.'[128]

The *phainomena* from which the principles are derived here are what is commonly agreed, but also what is observed. The dominance of everyday experience is marked throughout Aristotle's account of *phusis*. Any account whose consequences are at variance with what is normally perceived is automatically ruled out. Democritus, for example, who alone is praised for proceeding from 'arguments appropriate to the subject'[129] — that is, for proceeding from *archai* homogeneous with the subject matter of physics — is faulted on this count:

> '[Empedocles and Democritus], owing to their love for principles, fall into the attitude of men who undertake the defence of a position in argument. In the confidence that the principles are true they are ready to accept any consequence of their application. As though some principles did not require to be judged from their results, and particularly from their final issue! And that issue . . . in the knowledge of nature is the unimpeachable evidence of the senses as to each fact.'[130]

This kind of objection is all the more striking when we remember that the appropriate *archai* are derived from sense perception in the first place, since on Aristotle's accounts of *epagōgē* and *nous* there is really no other way they could be come by. The domain of evidence is the realm of sense perception: one arrives at principles on the basis of sense perception and the applications of these principles must agree with sense perception.

The effects of this highly restrictive feature of Aristotle's account can be demonstrated by examining his own 'proof' of the difference in essence between the earth and the celestial bodies, and of the non-generability, incorruptibility and immutability of the latter. Whereas terrestrial bodies

tend to move in a straight line, celestial ones revolve uniformly around the earth. Circular movement is a simple
movement,[131] and therefore there must be an element
associated with it. This new element is the ether. Ether has a
special status because the motion with which it is associated,
circular motion, is perfect in virtue of the fact that it is
complete in itself — whereas an unlimited straight line lacks
this completeness, and a limited one is necessarily imperfect
inasmuch as it has limits.[132]

A curious feature of the ether is that it is neither heavy
nor light; to make it heavy or light would be to confer on it a
tendency to move to its natural place, where it would of
necessity come to rest. It is therefore an element in a rather
strained sense, for the main principle by which the other four
elements are distinguished is the principle of levity and
gravity. Since circular motion can have no contraries, the
ether is exempt from generation, corruption, qualitative and
quantitative change: it undergoes only local motion.[133] We
are told that 'our theory seems to confirm experience and to
be confirmed by it'.[134] This remark is not an aside. If all
things have within themselves the *archai* of motion and rest,
why is it that some bodies can only undergo natural rectilinear motion and others only natural circular motion? The
answer can only be that we *see* the circular motion of the
celestial bodies and we *see* the rectilinear motion of
terrestrial bodies.

We must ask what claims this 'proof' can make to being a
demonstration of a difference in *essence* between celestial
and terrestrial bodies. The proof is not a demonstration in
the required sense, of course. We find no demonstrative
proof in Aristotle's physics and cosmology. The demonstrative syllogism is an unattainable ideal. The 'proofs' we are
offered are simply abstractions from sense perception and
these, by Aristotle's own criterion, are not scientific proofs.
The presentation of the results of abstraction as 'essences' is
quite implausible. Indeed, with new observations a surfeit of
new 'essences' soon begins to appear. Galileo's discovery
that the lunar surface is mountainous, for example, elicited
three new essences, on top of Aristotle's original celestial
substance, from Peripatetic physicists.[135] So much for the

claim that Aristotelian physics and cosmology are not cumulative!

§7   Conclusion

Aristotle presents us with a unified physics and cosmology which forms a highly articulate and closely coordinated system in close agreement with our sense experience. With Aristotle, physics and cosmology are henceforth serious and independent domains of study, as worthy of investigation as mathematics and metaphysics. This is a truly remarkable innovation. But what is achieved is achieved at a cost.

Aristotle puts forward an account of explanation which, could it be realised, would indeed lead to absolute certainty in our accounts of physical phenomena. Explanation would take the form of articulating the definitions of self-differentiating kinds, on the basis of *archai* construed as basic *aitia*, which would in turn open the way to deriving the essential properties of the phenomena which concern us. The trouble is that explanations of this kind cannot be given in principle. The major difficulty lies in the relation between the ontology and the evidential domain of the explanatory structure that Aristotle proposes we work within. The evidential domain consists simply of what sense experience tells us — as we have noted above, *endoxa* play a secondary role which is always subordinate to sense experience. The ultimate objects of *epistēmē*, for Aristotle, are sensible things and their relations. We have knowledge of a thing when we know its *aitia*. Now while it is possible that reference to the domain of evidence may establish the universality of *aitia* (in principle, but not in fact — at least in cases of interest in physics) it cannot establish their necessity. This is equivalent to saying that *aitia* cannot be established by reference to the domain of evidence proposed: that is, by *epagōgē*, with or without the inclusion of *nous*.

I am arguing that the explanatory structure that Aristotle proposes we operate with is incoherent, in that explanations of the kind required cannot be given in principle. There is a second point which I also want to stress. It is that this explanatory structure, while it does not generate explanations, does generate a particular kind of account of the phenomena under investigation. In the first place, it involves notions of what counts as evidence. It excludes certain kinds of situation as being evidential — for example, it denies evidential value to a situation to which we have no experiential access. Situations which we never have, do not, and never will perceive — such as bodies falling freely in a void — are excluded as being of no evidential relevance to physics, in particular. On the positive side, everyday experience becomes absolutely crucial not only in determining whether a theory is true, but also in delimiting what we must take into account in setting about demonstrating something.

Secondly, the account of explanation proposed by Aristotle proscribes the use of mathematical *archai* in physical demonstration.[136] This is a serious and well thought out move on Aristotle's part, and it represents a massive advance over the Pythagorean and Platonist alternatives, on which physics is simply reduced to mathematics on an explicitly mystical basis, or where true knowledge (*epistēmē*) of *phusis* is denied as being impossible in principle, and where *phusis* is located in the realm of opinions. Aristotle has faced up to a serious problem here, and its solution is not so easy as many commentators have thought. Galileo, as I shall argue later, treats it as a serious problem.

In connection with the question of mathematics, it is worth remembering that the kind of requirements which Aristotle imposes on explanation are derived, in the main, from the kind of requirements that had already been imposed on geometrical demonstration from the middle of the fifth century.[137] Aristotle used the notion of *proof* in this geometry as a model, adding the principle of homogeneity. The point of such a procedure was presumably that the certainty obtained in geometry — about which Aristotle had no doubts — could be achieved in physical reasoning also, provided physical reasoning could be conferred with a formal structure the same as that of a formalised geometry. In some ways, of course, this project is equivalent to Galileo's, but its proposed realisation is quite different. In order to determine in what way it was different, and what conditions need to be fulfilled before the project could be successfully realised there is a very important conceptual issue with which we must deal. I have argued that the uncritical reliance on sense experience and the rejection of the use of mathematics in physical proof are essential features of the account of explanation in physics which Aristotle proposes. In the following chapters, I want to examine the question of how the explanatory failure inherent in Aristotelian physics is to be dealt with. In the next chapter we shall look at attempts to develop and revise the Aristotelian account. I shall argue that the attempted revisions are inherently and necessarily unsuccessful, and that the whole explanatory structure of Aristotelian physics has to be rejected and replaced by a new one. This new one can only be successful if the questions of evidence and proof and, ultimately, ontology, are *re-posed*. Our discussion of the revisions to the Aristotelian account should enable us to determine more precisely what general kinds of question need to be asked if we are to be able to *explain* 'physical' phenomena.

Notes: Chapter 4

1   Cf. Łukasiewicz, *Aristotle's Syllogistic, passim*. Aristotle often uses temporally indefinite sentences — i.e., ones whose truth value may change — rather than atemporal sentences — i.e., ones whose truth value does not change — when giving examples. The former sentences cannot be construed as propositions in the formal sense. On this question cf. Hintikka, *Time and Necessity*, ch. 4.
2   *Topics,* 100a 18 ff.
3   *De Sophisticus Elenchis*, 164a20 ff.
4   *Posterior Analytics*, 71b18-25.
5   *Prior Analytics*, 53b8.
6   *Posterior Analytics*, 71b30-31.
7   *Ibid*, 72a30 ff.
8   *Ibid*, 73a20 and 74b5; see also *Nicomachean Ethics*, 1139b20.
9   *Posterior Analytics*, 73b16-18; see also 74b5 ff. It is clear that a theory of syllogisms with apodeictic premisses must allow for a distinction between universal and necessary propositions. This distinction is made at 73b25-28, where a necessary predication is characterised as being both universal and essential.
10 *Ibid*, 73b27-28.

11 *Metaphysics*, 1028b8-23.

12 *Ibid*, 1029a2 ff.

13 In chapters 10 and 11 of Book Z this 'primary substance' is identified with intelligible universals, which turn out to be 'one': God, who is pure form (cf. Book∧, 1074b15-1075a11) and hence identical with His essence. As Plotinus first noted (cf. Jaeger, *Aristotle*, p.351), it is obscure how we differentiate non-material entities, and hence it is obscure why God can be called 'one' at all.

14 *Metaphysics*, 1038b8-14.

15 Albritton, 'Forms of Particular Substance in Aristotle's Metaphysics'.

16 Woods, 'Problems in *Metaphysics* Z'.

17 Cf. *Metaphysics*, 1038a19.

18 For example, the individual substance 'Socrates' is a man and thus essentially mortal, but he is not essentially white; even though both mortality and whiteness can be predicated truly of Socrates. Socrates' mortality follows from the kind of thing he is, his whiteness does not.

19 Wieland, *Die Aristotelische Physik,* p.266 ff.

20 Cf. *Physics*, 195b17. There are occasional exceptions to this but none in the *Physics*. For details see Wieland, *op. cit.,* p.266.

21 On the non-sacrosanct nature of the classical conception of cause and effect see Anscombe, 'Causality and Determination', and Zilsel, 'The Genesis of the Concept of Physical Law'.

22 Carteron, *La Notion de Force dans le Système d'Aristote*, p.27.

23 Cf. Ross, *Aristotle's Metaphysics*, I, pp.126-7.

24 *Metaphysics*, 1013b 4-6.

25 Cf. Wieland, *op. cit.*, pp.173-187. Attributes had been converted into subjects by the time of Plato, functional concepts only by Aristotle. This conversion is rather important because Aristotle divides words into nouns (*onomata*), verbs (*rēmata*), and what can be called 'connectives' (*sundesmoi*). The third category includes everything not included in the first two: adverbs, adjectives, prepositions, conjunctives and (arguably) the article. Now only nouns and verbs have meaning in isolation — the *sundesmoi* do not, they only have a grammatical function. (On this issue see Robins, *Ancient and Medieval Grammatical Theory in Europe*, pp.19-20.) The conversion into subjects allows a correlative conversion of

parts of speech such as adjectives, prepositions and interrogative pronouns into nouns. For example, Aristotle, in constructing his categories, converts *posos* — 'how much?' — into *to poson* — 'quantity' or, literally, 'the how much'. This is not simply a grammatical manoeuvre, it is a procedure which has profound epistemological consequences, as we shall see later.

26 Owens, *The Doctrine of Being in the Aristotelian Metaphysics*, pp.130-1.

27 *De Interpretatione*, 16a3-4.

28 On this see Parts I and II of Wieland, *op. cit.*

29 Randall, *Aristotle*, ch.3.

30 *Posterior Analytics*, 100a10-b5.

31 *Prior Analytics*, 46a18-21.

32 *Metaphysics*, 980a21-27.

33 *Posterior Analytics*, 99b20 ff.

34 *Nicomachean Ethics*, 1145b2 ff.

35 On this see also *Topics*, 101b21-23.

36 *Ibid*, 101b3-4.

37 Cf. *ibid*, 163b9.

38 Cf. Hamlyn, *Aristotle's De Anima*.

39 *Topics*, 108b11.

40 The species 'man' constitutes a finite domain, for example, but we do not have access to every member of that domain.

41 *Posterior Analytics*, 87b30 and 92b39 respectively.

42 *Ibid*, 71b33-72a5.

43 *Physics*, 184a16-184b3.

44 Cf. Wieland, *op. cit.*, ch.1, esp pp.59-69.

45 Cf. Barnes, *Aristotle's Posterior Analytics*, p.254.

46 Randall, *op. cit.*, p.43.

47 *Posterior Analytics*, 100b13.

48 See, for example, Ross, *Prior and Posterior Analytics*, p.49. This interpretation has been questioned in Buchdahl, *Induction and Necessity in the Philosophy of Aristotle* (p.18 ff.), as well as in the articles of Kosman and Lescher cited below.

49 Cf. Kosman, 'Understanding, Insight and Explanation in Aristotle's *Posterior Analytics*', p.385.

50 Cf. *Posterior Analytics*, 84b35-36. These are the 'elements' referred to at the beginning of the *Physics* (cited above).

51 Lescher, 'The Meaning of Nous in the *Posterior Analytics*', p.64.

52 *Metaphysics*, Book E.

53 *Physics*, 193b32-35. Strictly speaking, Aristotle is concerned here with things which 'supervene on' bodies, rather than with

'attributes' or 'accidents' of bodies, but this does not affect the argument. Cf. Charlton, *Aristotle's Physics*, p.93.

54 *De Caelo*, 297a9 ff.

55 The account of *arithmos* that I shall present is based on the work of Klein (*Greek Mathematical Thought and the Origin of Algebra*). A detailed defence of the kind of interpretation that Klein puts forward can also be found in Szabó, *Anfänge der grieschischen Mathematik* (see especially the criticisms of Neugebauer and Van Der Waerden, pp.455-488). See also Mahoney, 'Babylonian Algebra: Form Vs Content' and 'Another Look at Greek Geometrical Analysis' for a discussion of related topics.

56 *Physics*, 224a1 ff.

57 *Republic*, 525d.

58 *Physics*, 221b14.

59 Cf. Plato, *Theaetetus*, 195d-196b and *Republic*, 526a; and Aristotle, *Posterior Analytics*, 76b4 ff.

60 *Metaphysics,* 987b14 ff. and 1090b35 ff. This account contains much more than we find in the extant work of Plato, and it ascribes to him a kind of number — the *arithmos mathēmatikos* — which is *ontologically* intermediary between the *arithmos eidetikos* and the *arithmos aisthētos*. I shall argue in §5 that while this interpretation is compatible with the account of 'mathematical numbers' in the *Timaeus*, it is not the only interpretation which is compatible with this account, as is usually thought.

61 The reasons for this are complex and to go into them here would require a full exposition of Plato's account, which would be out of place in the present context. The question is dealt with fully in Klein, *op. cit.,* pp.61-113. See also Szabó, *op. cit.*, p.79 ff.

62 Cf. *Metaphysics*, 1029b13 ff.

63 Cf. *ibid*, 1080a15; 1082a16; 1083b22.

64 Cf. *ibid*, 1077a15-18.

65 *Physics*, 219b5 ff and 220b4 ff.

66 *Metaphysics*, 1057a3 ff.

67 *Ibid*, 1016b21 ff.

68 *Ibid*, 1082b35.

69 *Ibid*, 1056b23 ff. I have used Klein's rendering of the Greek here (cf. Klein, *op. cit.*, p.108).

70 *Metaphysics*, 1016b17-20; 1021a12 ff; 1088a6-8.

71 Euclid, *Elements*, books 7-9 (Vol. 2, pp.277-426). For a critique of Heath's commentary on these books — in which he attempts

to present their content in terms of an algebraic formalism —
see Unguru, 'The History of Greek Mathematics', p.88 ff.

72 Cf. Klein, *op. cit.*, chs. 11 and 12.

73 *De Caelo*, 306a9-12; cf. also *Posterior Analytics*, 76a23 ff.

74 *Posterior Analytics*, 75a28-38.

75 *Ibid*, 75b16-18.

76 *Physics*, 193b34 ff.

77 The forms *simpliciter* with which the mathematician deals are
only *simpliciter* insofar as they are separable in thought, as we
have noted in §3.2. The mathematician does not deal with *pure
forms*: this is the domain of the metaphysician, who is concerned with being-*qua*-being (i.e. God, as defined in
Metaphysics, $\wedge$, 9).

78 Cf. *Metaphysics*, 995a14-16 and 997b34-998a6.

79 *Posterior Analytics*, 75b14 ff and 76a9 ff.

80 *Ibid*, 76a12.

81 *Ibid*, 76a15.

82 *Physics*, 194a10. The term 'optics' here denotes perspective
and catoptrics; it does not include the study of the nature of
light or visual perception. Cf. Lindberg, *Theories of Vision
from al-Kindi to Kepler*, pp.6-11.

83 *De Caelo*, 303a20-23.

84 Dk frag. 8, 1n. 50 ff; in Kirk and Raven, *Presocratic
Philosophers*, p.278.

85 We never find an argued rejection of Parmenides' denial of
change, even when Plato explicitly takes issue with Parmenides
over the question of not-being. See, for example, *Sophist*,
244b ff and *Parmenides*, 136a ff.

86 *Symposium*, 211b and *Phaedo*, 78d ff.

87 *Sophist*, 254c ff. The justification for this is that it makes the
blending of Forms possible.

88 *Timaeus*, 28a-c.

89 *Ibid*, 29a.

90 *Ibid*, 29c.

91 *Ibid*, 29c.

92 *Republic*, 527a. It is worth noting that Plato criticises the
'geometers' for thinking that their axioms are certain, whereas
they are really only hypothetical. On this see Robinson, *Plato's
Earlier Dialectic*, ch. 10.

93 *Timaeus*, 48e-53c; esp. 50c-d.

94 *Ibid*, 49d.

95 *Ibid*, 52d.

96 *Ibid*, 52d-53a.

97 For a detailed account of the relation between *genesis* and

kinēsis see Solmsen, *Aristotle's System of the Physical World*, ch.2.

98 *Timaeus*, 57d-58a.

99 Spherical rotation is the most perfect kind of motion because it combines movement and stability — the latter insofar as it involves no change of place (*Timaeus*, 40a). The regular celestial motions are the 'moving image' of that eternity which characterises true being.

100 On this see also *Laws*, 898a ff.

101 Cf. *Metaphysics*, 1025b20-1026a12.

102 Nevertheless, Aristotle takes Parmenides' argument as valid for the persisting substratum itself. This substratum can neither come into being nor cease to exist. Aristotle's argument for this (*Physics*, 192a25 ff) is similar to that of Parmenides. In the case of *metabolē*, or change in general, the doctrine of prima matter — which can be regarded as the ultimate substratum — comes very much to the fore. There is considerable disagreement on the status of prime matter. I would tend to accept Wieland's interpretation of this doctrine, on which prime matter is only a limiting concept which plays no essential role in Aristotle's philosophy (cf. *Die Aristotelische Physik*, p.209 ff). For the opposing interpretation see Fitzgerald, 'Matter in Nature' and Owens, 'Matter and Predication'. In the context of the *Physics*, we shall be restricting our attention to *kinēsis*, so the doctrine of prime matter will not concern us.

103 *Physics*, 192b8-23.

104 Solmsen, *op. cit.*, p.98.

105 Unfortunately, so too is the doctrine of the Unmoved Mover, which in the final analysis is also responsible for all movement and change. The Unmoved Mover is an *archē* of motion and is not to be confused with God. This becomes clear when we consider that Callipus' demonstration of the independence of the motions of heavenly bodies elicited a corresponding 55 Unmoved Movers from Aristotle in Book Λ of the *Metaphysics* (cf. Jaeger, *op. cit.*, ch. 14). There is surely some discrepancy between the doctrine of motion which finds its *archē* in the Unmoved Mover and the teleological conception of *phusis* in which everything has its own end, and in which *phusis* itself is an *archē* of motion, although Kosman has attempted to reconcile these two to some extent (see footnote 132, below).

106 *Physics*, 201a10-11.

107 Cf. *De Anima*, 417b10 ff.

108 *Physics*, 201b30.

109 *De Anima*, 417a21-30.

110 Kosman, 'Aristotle's Doctrine of Motion', p.51 ff.

111 *Physics*, 201b9-15.

112 *Metaphysics*, 1048b23-25. For a full discussion of this passage see Ackrill, 'Aristotle's Distinction between Energeia and Kinesis', pp.122-128. See also Aubenque, *Le Problème de l'Être chez Aristote*, pp.438-456.

113 *Physics*, 201b31.

114 Kosman, *op. cit.*, p.51 ff.

115 As was noted in ch. 1, only processes require explanation.

116 *Physics*, 212a5-7.

117 *Ibid*, 212a14-21.

118 Owen, 'Tithenai ta Phainomena', p.178.

119 For example, his distinction between proper and common place allows him to escape the consequences of Zeno's paradox of place, *viz*, everything is in place, and it is therefore in something, but place is something and therefore itself in something, and so on *ad infinitum*.

120 *De Caelo*, 285b16-17 and 287b22-288a12.

121 *Physics*, 255b8 ff.

122 *Ibid*, 225a11-13.

123 *Ibid*, 214b24 ff; also *De Caelo*, 300a1 ff. It is clear that once we have accepted the idea of natural motion (in the Aristotelian sense), the dynamical problem of falling bodies is not the same problem as that of classical mechanics. This is the source of some of Galileo's troubles. If the fall of a body is due to its natural tendency to reach its natural place, it is only to be expected that a body 'moves more quickly' as it approaches its end (*De Caelo*, 277a27 ff).

124 *Physics*, 215a5-9.

125 Cf. *Timaeus*, 79b.

126 Galileo, *Dialogo*, pp.236-7.

127 *De Caelo*, 311b16-18.

128 *Ibid*, 311b20-24.

129 *De Generatione et Corruptione*, 316a13.

130 *De Caelo*, 306a12-18.

131 The criteria for simplicity here are geometrical: the motion is made along lines all of whose parts are similar. The situation is complicated by the fact that this definition would not exclude a cylindrical helix, whereas a cylindrical helix is not acceptable to Aristotle as a simple geometrical figure, as is clear from *Physics*, 228b24-5.

132 As Kosman has noted, 'actuality is de-motionalised being not by virtue of having been brought to quiesence, but by virtue of having become entelic, having become its own end. It is for this

reason that circular motion, of which each part is as much the
end as any other, is the closest analogue to full reality. In one
sense that reality is most closely approximated in the axis of
such motion, in the still point at the centre of the most
energetic activity . . . In another sense it is equally the circumference which, in establishing limit in the very steadiness of
circuit and ever-regerenated newness, resembles that reality.
Aristotle's unmoved mover, as much at the circumference as at
the centre of the cosmos, descends from that more ancient god,
the great encircling Okeanos, forever flowing and nourishing,
yet unchanged.' ('Aristotle's Definition of Motion', pp.59-60).

133 *De Caelo*, 272a12 ff.

134 *Ibid*, 270b4-5.

135 Cf. Clavelin, *The Natural Philosophy of Galileo*, p.392.

136 I have noted above the qualifications which must be made to
this with regard to the 'subordinate sciences'. Particularly
interesting in this respect is the spurious, but nevertheless
Peripatetic, *Mechanica*. We find there an account which often
works in terms of arithmetical ratios. However, as Carteron
(who wrongly ascribes the work to Aristotle) has shown, the
study of the variations in ratio are most often proposed only in
order to examine the limiting cases — that is, the cases of zero
and infinity'— and there is nothing to suggest that their cor-
rectness in intermediate cases is being asserted, or even that the
author considers these to be of any interest (cf. *La Notion de
Force dans le Système d'Aristote*, p.27 ff). In the main, the
ratios are introduced simply as a way of expressing that which
is sensibly given in the light of such principles as 'the greater
corresponds to the greater' and so on. (cf. *ibid*, p.28)

137 For a detailed account of the parallels see Lee, 'Geometrical
Method and Aristotle's Account of First Principles'.

CHAPTER 5

PHYSICAL EXPLANATION AS SYLLOGISTIC DEMONSTRATION : II

§1 The Representation of Reality: Nature as *Signa Dei*

IN discussing Aristotle's account of commensurate universals, we noted that a distinction is made between the order of being and the order of knowing, and that this distinction is primarily one between different 'forms of knowledge'. For Western Christian philosophers from the Patristic period onwards this distinction is an ontological one of fundamental importance.[1] Within this philosophy, sense perception can conduce a subject to an accurate knowledge of the causal principles which constitute prior and non-sensible realities. These causal principles are really prior in the order of being but posterior in the order of knowing. Further, the order of being is not the order of things and their relations, in the Aristotelian sense, but the order of spiritual reality. It is spiritual reality which makes human thought possible. Hence the Being (that is, ultimately, the existence *and* essence) of God is not only a condition of whatever man might know about Him, it is also a condition of whatever man might know about anything. There is an intimate relation between God's Being and man's knowledge. As Colish notes, 'the relationship between man as knower and God as object of knowledge depended not only on God's absolute and transcendent existence. It also depended on His decisive intervention into the mutable world of His creation'.[2] Epistemologically, God is beyond the scope of human knowledge, but He makes such knowledge of Himself as He wishes to communicate generally accessible to man. For Christian thinkers this knowledge is communicated through human language, lost at Babel and reborn through the Incarnation.

The sensory 'data' which for Aristotle formed the starting point of knowledge are transformed into verbal signs in the Medieval account of knowing, and Aristotle's *organon* is thereby replaced by the Medieval *trivium*: rhetoric (speaking and literature), grammar (reading and language), and dialectic (thinking and logic). The epistemology of the *trivium* is symbolic and is cast in terms of the realm of words and the realm of reality. Language is the cognitive intermediary between the knower and the object of knowledge. Words, in turn, may signify truly, if partially, that which really exists.

In this epistemology, the accuracy of a verbal formulation depends on its correspondence to the object it seeks to describe. Statements are not therefore productive of knowledge in the first instance, but expressive of a knowledge which must already exist in the mind of the knower. This means that the subject must have an anterior knowledge of the object, a knowledge indispensable to his recognition of the truth of the words expressing it. Hence we can distinguish two functions of language: the indicative function and the commemorative function. In its indicative function, language can point to the object if it is not already directly known. Commemoratively, language can enable the subject to recall the knowledge of an object which had previously been introduced into his mind at some specific point by the object itself. This doctrine clearly has Platonist origins, but the commemorative function of language should not be equated with Plato's doctrine of recollection,[3] where knowledge has *always* been in the mind. It is similar to the evocative function conferred on icons in Byzantine Iconic Epistemology, but this latter only allows the sign a commemorative function.[4]

The thesis common to the Medieval Christian philosophers from Augustine to Aquinas is the thesis of words as signs in the knowledge of God. They all employ a theory of signification in which a fundamental duality — being/knowing — is posited, where being is (logically and chronologically) prior to knowing. This theory of signification is expressed through the three modes of the *trivium* in Augustine's rhetoric, Anselm's grammar, and Aquinas' dialectic.[5]

For Augustine, words are acoustic sounds which accurately but partially correspond to the realities they represent. Words have a special status as signs insofar as they are the only signs which can represent themselves. They can be used to make false statements, of course, but this occurs only when the speaker deliberately perverts their true signification, or when he is inept in expression.[6] Words signify really existing things either commemoratively or indicatively in the subject's mind, depending on his previous relation to the object. However, verbal signs cannot be cognitive in the first instance: they must be 'energized' in the mind of the knower in order for them to conduce to the knowledge of their significata.[7] These conditions obtain whether the words cause the subject to know or whether they serve to deepen his understanding.

Anselm's main epistemological concern, like that of Augustine, is the theological problem of speaking about God. But, as Colish has noted, 'where Augustine sees the task of theology as the eloquent expression of the Word, Anselm sees it as the conscientious and faithful definition of the Word'.[8] Anselm distinguishes between those signs (in this case words) which have rectitude and those which only have truth. A statement is accidentally true — it has truth — if it expresses the speaker's meaning; it is naturally true — it has rectitude — if it also expresses realities accurately. Discourse would have no rectitude unless its objects existed, and these objects exist because God creates and sustains them. Anselm equates rectitude and necessity, and claims that an argument may be considered to have been proved by necessary reasons when it states that things are the way they really are. Words and ideas are controlled and validated by the objective realities they are designed to express, and they can express these realities, in their own terms, with rectitude. The theory of signification which Anselm puts forward in the grammatical mode is substantially the same as that which Augustine had put forward in the rhetorical mode: the accurate but limited status of signs in relation to their objects, and the different subjective functions of signs, depending on the cognitive and affective status[9] of their hearers *vis-à-vis* God.

Whereas Augustine and Anselm are concerned almost exclusively with knowledge of God and the soul, Aquinas also deals in detail with the 'natural' — as opposed to the 'spiritual' — world. His treatment of this is based on the same theory of signification. This theory of signification is translated into the mode of logic with Aquinas,[10] and the primary terms of the analysis are now ideas, on the one hand, and the created universe as the principal set of *signa Dei* on the other. Signs are still primarily verbal but they no longer have to be articulated vocally; the idea or concept is a *verbum mentis*. This move from words as acoustic signs to words as ideas which may or may not be enunciated acoustically finds its initial foundation in the supposition theory of William of Sherwood (or Shyreswood) and Peter of Spain.[11] A term has 'supposition' when it replaces or stands for whatever it is intended to signify in a proposition. It has 'signification' insofar as it is an arbitrary sign which has the function of indicating some object; when this function of signification is actually *exercised* then the term has 'supposition'. Whereas a theory of signification merely studies the sign-relation of terms in general, the theory of supposition studies the signs or terms as predicates in relation to their subject or subjects. Inasmuch as the relations between signs and their significata are arbitrary and conventional they are automatically accurate. Supposita, on the other hand, involve intellectual judgements of truth; the term 'supposition' implies that the subject knows what he means by a word when he uses it in a proposition, and that he can form a judgement about it.

Whereas Aristotle's logic had been presented in terms of formulas containing variables replaceable by 'object language' expressions, later Medieval logic — that is, the *logica moderna* initiated by William of Sherwood and Peter of Spain — was formulated metalinguistically by means of rules referring to language expressions. More importantly, there was a crucial vagueness in the earlier accounts of predication (notably those of Aristotle and Boethius), in that the status of predicates was never made clear, and it was often suggested that they might be extralinguistic or extramental entities. The *logica moderna* is explicitly a science of linguistic entities only. Indeed, it is a central feature of supposition that it is not a relation between a term and the designatum of that term, but a relation which one term of a proposition has to another, this relation being that of having some part of their extensional domains in common.[12]

The distinction between signification and supposition is similar to the Anselmian distinction between truth and rectitude. However, whereas the Anselmian judgement of rectitude starts with the mind of the speaker and moves to the acknowledgement that his words correspond to an 'exterior' reality, the step from signification to supposita indicates that a real thing has entered the mind of the subject by means of its species, that the subject has come to know what the sign represents, and that he is capable of relating it to other things that he knows.[13] It is a short step from this theory to the Thomist identification of signs as concepts, whereby the word is signified in the mind by ideas which are formed from the intelligible species of things, and words are required to signify these things only insofar as one wishes to communicate them to others.

Thomist epistemology is distinct from that of Augustine and Anselm in two crucial programmatic respects: in Aquinas' concentration on nature as the principal source of signs of God, and in his concentration on the question of how knowledge enters the mind rather than on the question of how statements can be objectively true. Aquinas' account of the second problem, taken independently, is clearly parasitic on Aristotle's account of perception in the *De Anima*. The Thomist treatment of perception is explicitly epistemological, however, in a way that Aristotle's is not. Like Aristotle, Aquinas considers perception as a type of change in which the sense organ is altered. But he also posits a spiritual change resulting in a *phantasma* — a particular mental entity — corresponding to a physical change in the sense organ. Now although it is the 'passive reason' which is the source of our knowledge of intelligible forms or species, this passive reason cannot function without the 'active reason' since, as with the acquisition of sensible form, the acquisition of intelligible form is a process in which something potential is made actual. What is made actual in this case is knowledge. The function of the active reason is to abstract the form or *species impressa* from the *phantasma*. When imposed on the passive reason, the *species impressa* produces a *species expressa* or *verbum*.[14] The outcome of this process is a *concept* which has a verbal form, even if this concept is only expressed internally and not articulated acoustically.

This process can be described at another level. In the Middle Ages, a distinction is made between the *vegetative soul*, which is responsible for such things as growth; the *animal soul*, which is responsible for locomotion and 'apprehension'; and the *rational soul*, which is concerned with man's immortality and his ability to think. The faculty of apprehension is divided into the *external* and the *internal* senses.[15] It is by the external senses that we know objects directly: these objects act upon the senses by means of intelligible species or forms. These external senses each have their proper objects — the proper object of vision is the visible, the proper object of hearing is the audible, and so on. Because of this, the external senses can be subdivided into the five 'proper' senses. The external senses are 'passive' inasmuch as they require an outside agency — a 'species' or similitude — if their potentiality is to be actualised.

The internal senses can only know an object via the mediation of the external senses. The five powers of the internal senses are common sense, imagination, estimation, phantasy and memory. There is a potential knowledge in each of these powers which a species, which has acted on the external senses, actualises. This actualisation is successive, and the five powers correspond to five consecutive processes. The function of the *common sense* is to connect the species received in the proper senses. The *imagination* stores the image (*phantasma*) of the sensible object known through its species. The function of the *estimation* is the perception of species which are sensed by neither the proper senses nor the common sense. These species are called 'intentions', and they include such things as friendliness and hostility. The function of the *phantasy* (or 'cogitative power') is to connect the species of the common sense and those of the imagination (i.e. the 'intentions'). Not all Medieval philosophers distinguish the estimation and the phantasy: Albertus Magnus (following Avicenna) does, but Aquinas (following Averroes) does not. The last of the internal senses is *memory*, which stores the species and intentions received in the other internal senses for future reference. Finally, the rational soul abstracts the universals of intellectual cognition from the internal senses.

Now it might be considered that this kind of account — especially in its Thomist version, where it takes the form of a *De Anima* commentary — simply makes explicit much that is already implicit in Aristotle's work. In particular, the Thomist account of the relation between the rational and the animal soul bears a striking similarity to some interpretations of Aristotle's discussion of *nous*. After all, all we are really left with in Aristotle's account, construed epistemologically, is abstractionism. Aristotle tries to avoid abstractionism by the introduction of *nous*, but this introduction is unsuccessful, as we have already noted in the last chapter. He operates with abstractionist procedures in practice and this precludes explanations of the required kind being given in areas such as physics. But Aquinas is neither simply bringing this problem to light, nor simply accepting abstractionism as the only procedure by which knowledge is to be attained. His 'revision' of the Aristotelian account is made on the basis of a theory of signification which is alien to Aristotelian epistemology. The ultimate object of 'knowledge', for Aquinas, is God: this is the most important consequence of the doctrine that natural things are the principal *signa Dei*. On this account special problems arise when we deal with those things which are not sensible, that is, the things devoid of matter, such as universal causes. Aquinas proposes a solution to this problem in terms of a particular ontological interpretation of the original Aristotelian distinction between the orders of being and knowing. He argues that universal causes are the most intrinsically intelligible things in the universe, because they are the most actual. But because they are not sensible, they are the least well known. There is no paradox involved here. We must remember that Aquinas is a thorough-going realist, and that on the Thomist account of perception the form of the thing perceived actually enters the mind of the perceiver. Intelligibility is a property of things in themselves for Aquinas, it is not a property which things acquire only when they exist in the mind.

On the Thomist account, the world as it exists in the mind of God is prior to the world as we know it. It is ontologically prior because God is the reason why the world, which is the sign of God, exists; God can exist without the world but the world cannot exist without God. It is also epistemologically prior — on a strict interpretation — in that we cannot have perfect knowledge of the world without knowledge of its cause or reason, God. Although Aquinas considers knowledge of the universe known through its causes to be accessible to man, such knowledge is not perfect: it is true but partial. God only reveals to us what He wishes us to know. If there were such a thing as ultimate knowledge of the universe it would be knowledge of God,[16] but there is a point where the knowledge of God by reason alone must come to an end; from there on we can only proceed via the gift of faith. This means that physical knowledge, insofar as it is possible, is subordinate to 'metaphysical' knowledge, and, in the final analysis, to faith.

There are two ways in which this subordination can operate. In the first place, physics can simply be subsumed under theology. This certainly happened in the Middle Ages but the accounts which resulted are without interest for us since they rest, in the main, simply on Biblical exegesis or on commentaries to such works as Peter Lombard's *Senteniae*. (It is worth noting, perhaps, that a subordination of this kind is in complete contradistinction to Aristotle's account, on which the distinct subject genera of physics and theology require distinct *archai*.) The complete subordination of physics to theology is required on this strict interpretation of the *signa Dei* doctrine since in understanding nature as a sign one must understand what it signifies and how this is signified.

The alternative to this is to treat physics as an autonomous domain of study. Because of the *signa Dei* doctrine this treatment cannot be justified on quite the same basis as that provided by Aristotle. What has to be established in this justification is the idea that sensible things constitute the object of study of physics. That is to say, the physicist studies sensible things as such, and in this way neither infringes upon nor relies upon the theologian's account of things *qua* signs of God. The physicist studies nature *qua* the totality of sensible things whereas the metaphysician or theologian studies nature *qua* signs of God. We shall see later that necessary reference to God is made in the Medieval (and Renaissance) revisions of *nous* — that is, in establishing a link between what is being explained and the evidence for the universality and necessity of explanation. For the moment, however, let us restrict our attention to the question of the autonomy of physics, and in particular to its epistemic status.

Although the *Signa Dei* doctrine is common to all the Medieval Christian thinkers, it is a theory of signification and not an epistemology as such. It raises very specific epistemological problems, as we have seen, but it does not provide a solution to these problems. In fact, we can distinguish two main kinds of solution to these epistemological problems, both of which have an important bearing on the question of scientific demonstration. The first, which is the Thomist solution, introduces an account of perception in which the forms of natural objects are impressed upon our sensory faculties; these forms, or 'species', then provide the basis on which the rational soul proceeds to discover universal and necessary principles. The second kind of solution — that proposed by Ockham — is radically different. Whereas for Aquinas the process of cognition ends with the abstraction of an essence, for Ockham it ends with identifying likenesses which are completely, and indeed ontologically, inseparable from the individuals themselves. Ockham does not attempt to explain individuals in terms of universals. He treats universals solely as concepts, and his analysis of these concepts is linguistic.

Following Duns Scotus, Ockham admits two kinds of cognition: 'abstractive' and 'intuitive'.[17] On the Scotist account of this distinction, abstractive cognition is cognition of an individual in an intelligible species — that is, a 'diminished likeness' stopping short of the individual itself. Intuitive cognition, on the other hand, is direct cognition: it is cognition of individual existence. Ockham dispenses with intelligible species altogether and all (non-discursive) cognition is, for him, direct, whether it be intuitively of individuals or abstractly of their representations in the intellect. Although intuitive cognition provides us with knowledge of existence, its *object* is the same as the object of abstractive knowledge, *viz* terms.

In order to examine this theory more closely, we must return to supposition theory, which we have mentioned briefly above. The most generally influential exposition of supposition theory was Peter of Spain's *Summulae Logicales* (1246), which was the main dialectic textbook up to the third decade of the sixteenth century. The prime purpose of the *Summulae* is instruction in the selection and manipulation of existing statements above the world. This procedure has an important basis in the *trivium*, and especially in the relation between grammar and dialectic as these are conceived in the *trivium*. The fundamental concern in the study of grammar was the correspondence between the structure of language and the structure of the mental and physical worlds. As Jardine has noted, this conception of grammar contributes to 'a view of language as providing a perfect map for process and change', and dialectic is 'then seen as analysing natural relations as embodied in discourse, and manipulating language to gain insight into the natural world'.[18]

At the root of this view is an assumption of a pre-existing harmony between concepts and reality. Ockham rejects this assumption. He also rejects Peter of Spain's treatment of terms as having a 'natural' supposition.[19] In Ockham's theory of supposition, terms have supposition only when they are used in propositions, and they have proper supposition only when they are used with their literal meaning. Supposition is a function of a term which is either a subject or a predicate in a proposition. Terms have three modes of existence: as *mental* terms (concepts), which represent or make known what they signify; as *spoken* terms, which are arbitrarily — or, at least, conventionally — instituted to call to our mind the same thing that the mental term signifies; and as *written* terms, which are similarly conventional. (For a schematic representation of the main features of Ockham's theory of supposition see the Appendix to this chapter.)

Supposition theory is particularly important in Ockham's work because his theory that knowledge of things is knowledge by mediation of the supposition of terms. He argues that 'it is to be known that any science whatever, whether it be a real or a rational science [ie., one of the "speculative" sciences such as physics, or a science such as logic], is only of propositions as that which is known, because it is only propositions which are known'.[20] What is rejected in this statement is the idea that physics and logic are ontologically distinct in that one deals with things and the other with propositions. All knowledge is knowledge of propositions for Ockham. The distinction between logic and physics lies in the fact that the terms of the former stand for other terms, whereas the terms of the latter stand for things. Hence in physics what we 'know' *strictly* are terms (in propositions); knowledge is only *mediately* of things. Physics is a rational discourse — a body of propositions — concerning the natural world, but it is in no sense isomorphic with the natural world.

Now although intuitive knowledge is not knowledge of particular things as such, it is to be distinguished from abstractive knowledge by the fact that in intuitive knowledge we know the existence of what the terms signify. Such knowledge is generalisable but it does not lead to knowledge of essences. It is one of the central theses of Ockhamist epistemology that nature is inherently contingent: there is an absence of necessity in everything created. Indeed, because nature is a sign of God, and because it is supernaturally possible for an omnipotent God to intervene directly in the human mind without the mediation of His natural signs, intuitive cognition does not *necessarily* require the existence of its object. Nicholas of Autrecourt interprets this argument as providing grounds for scepticism, by arguing that if we admit that an effect can be produced supernaturally without its natural causes, then we have no right to posit natural causes for any effect whatsoever. For Ockham himself, and

Bernard of Arezzo following him, it is *only* when the intuitive cognition is caused supernaturally that its existence does not require the existence of its object. We can judge that an object is present before us by the principles of *natural* causation even though it is supernaturally possible for God to produce such an intuitive cognition in us without the mediation of the object. As a consequence, Ockham 'admits an order of natural evidence and necessity *secundum quid* or *ex suppositione naturae*'.[21]

Ockham does not consider that the propositions of the natural sciences are anything more than means of ordering concepts (i.e. mental *terms*) which 'supposit' real things. The reasons why he puts forward this thesis are of some interest, and they centre around the question of evidence. For Ockham, the only kind of evidence we can have concerning natural things is evidence of their individual existence. Since all existence is contingent the knowledge of natural things cannot be knowledge of something that is necessary, nor can it be used to infer general causes or principles from effects.

Ockham takes the domain of evidence of the physical sciences as his starting point, and in working towards the kinds of principles that could receive verification from this domain of evidence he concludes that such principles cannot represent any natural necessity — they cannot be a statement of essences. In doing this he locates one of the fundamental problems in the Aristotelian concept of explanation. However, this is essentially a negative advance. It serves to weaken the explanatory structure of Aristotelian physics, but it also reduces physical enquiry to abstraction from experience. Indeed, I hope to show later in this chapter that it is this restriction of the domain of evidence of physics which, by restricting the procedures by which physical concepts can be generated, precludes the question of explanation in physics being posed in such a way as to break the deadlock of Aristotelian physics. Since I shall elaborate on this point later, we shall not consider it further here.

The second issue that arises from Ockham's work concerns the aims of scientific enquiry. Physics, for Ockham as for Aquinas, is part of a more general project which is explicitly metaphysical and theological. Whereas Aquinas considers that physical knowledge is knowledge of real essences, however, Ockham rejects this conception of physical knowledge and substitutes for it an account of physical knowledge in which theories are simply devices for grouping phenomena in the simplest way. That is, providing the terms in the propositions which go to form a theory supposit really existing things, then a theory is adequate if it is the simplest thesis that suffices to save all the phenomena (*'sufficit ad salvandum omnia apparentia'*).[22] In defending this account of the status of physical theory, Ockham avoids the problem of how observation can reveal essential principles: a problem which Aristotle was demonstrably unable to solve, and which other Medieval (and, as we shall see, Renaissance) thinkers treated by recourse to Divine illumination. Nevertheless, Ockham's 'solution', while it is more rational and consistent than the alternatives, is proposed at great cost. Indeed, in one sense it constitutes a step back from Aristotle into a conception of physics which makes physical enquiry epistemically inferior to metaphysics or theology. Plato had conceived physics to be epistemically inferior to metaphysics, and in discussing his account I argued that part of the value of Aristotle's project was the attempt to render physical enquiry an independent domain of investigation in which physical theories enjoy a certainty comparable with mathematics and theology.

The final issue that must be mentioned here concerns the use of logical devices in physical theory. We have already remarked that, for Ockham, both logic and physics concern themselves with the analysis of terms in propositions;[23] they differ in the kinds of supposition that these terms have. This raises some major questions about the relation between physics and logic and the importance of these questions extends far beyond Ockham's writings. There are no straightforwardly Ockhamist thinkers in Medieval physics after Ockham, but there are many whose work is dominated by particular Ockhamist distinctions. Of especial interest in this respect is the kinematics of the Merton School, which raises problems concerning the relation between logic and physics in a particularly clear form, and is hence of some importance if we wish to elucidate the concept of explanation that is operative in Medieval physics.

§2 Logic and Kinematics

Up to now in this chapter, I have tried to describe some of the theoretical innovations which were introduced in Medieval theories of logic and language, and of signification in general. We shall now examine the way in which these innovations affect the explanatory structure of Medieval physics. The major theoretical problem of linking sensory evidence to universal and necessary explanations — a problem inherited from Aristotle, but treated in a different way by the Medievals — is something which we shall leave until later. With the exception of the treatment of this issue, Medieval physics is essentially a continuation of Aristotelian physics. Revisions and additions are made, but all of these are in keeping with the fundamental Aristotelian concept of *kinēsis*: indeed, they allow a tightening up of this concept. The Aristotelian conception of demonstration, and the account of explanation of which it is part, permeate Medieval discussions of physical problems. At the level of proof, mathematics is not endowed with an explanatory function although it is sometimes employed as a descriptive device — usually in astronomy, sometimes in optics but rarely in mechanics — in keeping with the Aristotelian account of the 'subordinate sciences'. At the level of evidence, sense perception and everyday experience (especially such things as the experience of muscular exertion) not only determine the criteria by which the truth and falsity of theories is established, but also determine the conditions under which physical concepts are formulated. It is from sense perception and everyday experience that the concepts of Medieval physics are ultimately derived. This does not necessarily restrict the domain of physical investigation to everyday experience, but it does determine how this domain, whatever its extension, is to be dealt with.

In considering Medieval and Renaissance treatments of both physical problems and problems in the theory of demonstration I hope to establish that the kind of difficulties which make Aristotelian physics unworkable are fundamentally the same as those which make Medieval and Renaissance physics unworkable. These problems are not simply ones of consistency or applicability. Great progress is made in both these respects in Medieval physics, but the fundamental problems remain. In order that we might examine the nature of these fundamental problems it is necessary that we present an account of some of the developments in physics from the thirteenth to the sixteenth centuries, but this account is not designed to compete with the many histories of the period which are available. It is simply a means to an end, the end being an assessment of the notions of explanation, evidence and proof with which Medieval and Renaissance mechanics operate.

In the mid-thirteenth century, Medieval mechanics was dominated by three problems: the proportionality of motive power (or 'force') and speed,[24] uniform motion, and the distinction between natural and violent motions. While these problems continued to be treated in the fourteenth century, the work of the Merton School in the first part of that century resulted in a change in the way the first and second problems were dealt with. The work of the Merton School has three distinctive features: (1) the rudimentary separation of kinematics from dynamics, and the subsequent establishment of the validity of treating motion in terms of its effects; (2) the attempt to describe the variation of qualitative forms in terms of degrees of magnitude; (3) the attempt to deal with non-uniform motions.

In the writings of the Merton School we find an explicit distinction between two kinds of account of motion: those which introduce the cause of motion and those which treat motion wholly in terms of its effects. Thomas Bradwardine, for example, in his *Tractatus de Proportionibus* of 1328, deals separately with the proportions of speeds of motion as related to the 'forces' of the moving and the moved body — motion *sub specie causae motivae* — and the same proportion related to the size of the moving body and the distance traversed — motion *sub specie effectus*:

'Having completed a general treatment of the proportion between the speeds with which motions take place with respect to both moving and resisting powers, it remains to demonstrate certain properties peculiar to the proportion of speeds in circular motions with respect to the quantities both of motion and of the interval traversed.'[25]

The distinction between motion *sub specie causae motivae* (the object of dynamics) and motion *sub specie effectus* (the object of kinematics) enables Bradwardine to determine the proportions between the speeds of bodies both with respect to their causes and with respect to their effects. Before we look at how he does this, we must examine the basis of the distinction.

The *conceptual* distinction between a thing's being in motion and its being moved is an innovation due to Ockham. First of all, for Ockham and the Merton School, as for Aristotle, motion itself is *not* an independent object of investigation. Motion is something which requires a substratum. However, Aristotle's account of *kinēsis* contained the anomaly that whereas changes in the quality and quantity of a thing were due to internal factors, change of place *seemed* to be due to external factors. This anomaly was cleared up by Duns Scotus who introduced the idea of the *ubi* (literally, the 'where') of a body.[26] On this theory, as a body moves from place to place a series of constantly renewed *ubi* are actualised. This idea, as well as strengthening Aristotle's account by unifying all the various forms of *kinēsis*, allows local motion, considered as a process occurring successively from one *ubi* to another, to be treated in terms of a succession of intensive magnitudes:[27] this is one of the central features of the Merton account, as we shall see below.

Secondly, the distinction Ockham proposes, and the Merton School follow, is a *conceptual* one. No *substantial* distinction is being made between a thing's being in motion and its being moved: Aristotle's dictum that all motion requires a mover is not being questioned. For this to be possible, the idea that all motions are processes would have to be discarded, and a law of inertia would be required.

Thirdly, on Aristotle's account the *conceptual* distinction is not made. For him, motion produced and the production of motion are the same thing. This is not to deny that

Aristotle recognised that the magnitude of a motion can be determined by distance traversed as well as by the motive power responsible for the motion. Indeed, in determining the speeds of two bodies in uniform motion he appeals precisely to distance traversed.[28] But, for Aristotle, in measuring the magnitude of a motion in terms of distance traversed one is measuring the *same thing* as when one determines the motive power responsible for the motion. Speed, in particular, is not considered as a characteristic dimension of motion; rather, it is a general feature of the magnitude of motions, something by which to compare motions.[29] This comparison is then simply reflected back onto the determination of the degree of magnitude of motive power. For example, we are told that the weight of bodies — as this is expressed in their physical constitution — is the *aition* of their (natural) motion. At the level of degrees of magnitude, this is expressed in the formula: 'as two weights are related to each other, so inversely are the times they take to fall'.[30]

Avempace and Aquinas had defined motion — in terms of its effects — as a traversal of extended magnitude in time, and they had construed this as being equivalent to a definition of motive power or force.[31] Ockham accepts the first definition but denies that it is equivalent to a definition of motive force. In fact, he accepts the more traditional account of Aristotle and Averroes on the question of force — as that which acts against a materially resistant body — but what is important is his separation of 'being acted upon by a force' from 'being in motion'. It is this conceptual separation which the Merton School immediately follows up, as we have noted above. We must now examine what they do with it. Let us look first at the treatment of motion in terms of its effects.

We have already mentioned the distinction between 'extensive magnitudes' and 'intensive magnitudes'. The former are those magnitudes which are amenable to direct measurement, whereas the latter are not amenable to direct measurement. For Aristotle himself, the first category includes distance and weight, the latter such things as hot and cold, dry and wet. Both kinds of magnitude are subject to increase and decrease. In the former, increase and decrease is by addition or subtraction of parts. Increases and decreases in intensive magnitude, on the other hand, are conceived in terms of *intensification* and *remission* of qualities.

In the work of the Merton School, speed is treated as an intensive magnitude — it is defined as *intensio motus* — subject to intensification and remission. This brings motion (*kinēsis*) *qua* change of place into line with motion (*kinēsis*) *qua* change of quality or quantity, therewith providing the possibility of a uniform treatment of *all* kinds of change which involve the *state* of a substance, as opposed to a change of the substance itself.[32] That is, it allows the uniform treatment of changes construed as occurring between contraries, as opposed to changes construed as occurring between contradictories (such as generation and corruption). By treating speed in terms of an intensive quality, changes of speed are now conferred with a precise meaning, and this represents a substantial improvement on Aristotle's own account. Change of speed is conceptualised in terms of *latitudo*, a concept which was originally introduced in the thirteenth century to characterise the qualities of a subject in terms of their capacity to change, and hence to distinguish these from the more substantial forms of a subject.[33] However, in the Merton usage, *latitudo* refers to the *process* of change and it usually takes a qualitative form. Thus Heytesbury, in his *De Motu*, refers to a body which, in equal parts of time, 'equalem acquirit latitudinem velocitatis' — acquires equal increments (*latitudinem*) of velocity.[34] The idea of *latitudo* as an increment provides the basis for a *description* and *classification* of the various types of motion depending on whether these increments increase or decrease, and the way in which they increase or decrease. For example, motions without any variation in speed are termed 'uniform'; motions with equal changes in speed in equal times are termed 'uniformly difform'; motions with unequal changes in speed in equal times are termed 'difformly difform'. Within this latter category, there is a further distinction between those motions with changes in speed subject to uniform variation — 'uniformly difformly difform motions' — and those with changes in speed subject to no uniform variation — 'nonuniformly difformly difform motions'.

The manner in which motions are classified and conceived by the Merton School throws much light on how we are to interpret their mean-degree theorem, or, as it is sometimes called, the Merton Rule. This Rule is usually expressed in modern notation, in which it states that for a body with an initial speed v_0 and a final speed v_1, the distance s traversed in time t is

$$s = \left[v_0 + \underline{v_1 - v_0}\right].t$$
$$2$$

There are several points to be noted about this Rule. The first concerns its expression in modern notation. The algebraic and functional notation in which we have expressed the Rule was not available in the fourteenth century. Relations between physical magnitudes continued to be given in the form of verbally expressed proportions, written at best as an equality of two ratios, right up to the eighteenth century.[35] This is a problem that we shall deal with later, but we can note now that such a representation does not allow the setting up of general equations since it does not allow for the distinction between an unknown variable and a parameter to be made in a satisfactory manner. There is also a more substantial issue at stake. In classical mechanics, the speed of any point is given in terms of a mathematical ratio defined by a certain operation, and nothing else. The Merton Rule is conceptually quite different from this. Fundamentally, what the Mertonians developed was not a *mathematics of motion* but a *language of motion*. For one thing, they simply did not have the appropriate mathematical concepts which would be required to give such an account.[36] Further, the new language of local motion was a language not *just* of local motion, but also of hotness and coldness, rarefaction and condensation, and so on. Local motion has no privileged position in this account: what binds these topics together is the fact that they are all changes occurring within a subject which itself does not change.

Secondly, the Merton Rule in no way contradicts the idea of motion as a finite process within set limits. Indeed, by making speed an intensive magnitude, the Merton treatment of motion actually strengthens the conceptual unity of *kinēsis*, since local motion can now be made subject to the action of a quality, like the other kinds of *kinēsis*, in a more rigorous and consistent fashion. In its most general form, the mean degree theorem states that a uniformly difform change of quality corresponds to the mean degree of that changing quality between a *terminus ad quem* and a *terminus a quo*. Hence, at the most fundamental level, motion is conceived in a completely Peripatetic way by the Mertonians. The Aristotelian doctrines of natural and violent motion, and circular and rectilinear motion, are affected in no way by the mean degree theorem.

Thirdly, where the work of the Merton School does differ from the Aristotelian account of motion is in the completely changed description of uniform and non-uniform motion. What exactly is involved in this redescription needs careful examination. The issue is best approached via Aristotle. Shapere has claimed that:

> 'The very words, uni*form* and dif*form*, reveal the connection with the general Aristotelian doctrine of change; in difform motion, a further changing form is superimposed on the original one.'[37]

As it stands, this statement is a *non-sequitur*, but a very interesting one: First, let us consider its logic. as we noted in the last chapter, it is an essential feature of the Aristotelian account of motion (and change in general) that neither the *terminus ad quem* not the *terminus a quo* themselves be motions (or changes). If we consider Aristotle's account of *kinēsis*, for example, it is clear that the reasoning behind this is sound. Since *kinēsis* is a process by which an object changes its properties, it is characterised in terms of the two contrary states between which the process occurs: the state of not having a particular property and the state of having this property. *Kinēsis* is a process which a persisting subject undergoes. The superimposition which Shapere speaks of is quite incoherent in Aristotelian terms since it would amount to the motion of a motion which, in turn, would mean that the termini of motion would themselves be motions. So if the first part of Shapere's sentence is correct the second part cannot be, and vice versa.

The situation is complicated, however, by the fact that there are several different kinds of account of intensive magnitude to be found in the thirteenth and fourteenth centuries.[38] The first main account follows a thesis suggested in Aristotle's *Categories*,[39] according to which qualities themselves are invariable: apparent changes in the intensity of a quality are due to a greater or less 'participation' of the subject in the quality. This account is the traditional one, accepted by Porphyry and Simplicius, and later by Aquinas and Aegidius Romanus. One important feature of the account is the construal of intensive and extensive qualities as different genera with correspondingly different modes of change or variation. Extensive magnitudes are subject to increase and decrease by addition and subtraction. Intensive magnitudes, on the other hand, are subject to increase and decrease by passing through a series of discrete and, at each instant, wholly new qualities. Here, the quality at a particular instant is wholly destroyed before a new one comes into being. Because of this, increase in intensive magnitudes is usually construed by analogy with the lengthening of days in summer, which does not occur by addition but by a succession of increasingly longer days.[40] The crucial claims of this kind of account are (1) that a quality cannot exist without inhering in a subject and since given individual qualities cannot migrate from one subject to another there is no basis for considering different distributions of the same quantities of qualities into greater or lesser extensions, and hence there is no physical basis for the use of quantities of qualities; (2) that qualities do not have parts.[41]

The second kind of account was held in many forms but its most important and influential version is that proposed by Gottfried of Fontaines and followed up by Duns Scotus. On this 'Scotist' account, the two genera of intensive and extensive qualities are merged, and intensification and remission are treated in terms of addition and subtraction. This is achieved on the basis of a distinction between the form and matter of any *quality*, the form of a quality being that in virtue of which it is a particular quality, and the matter determining the extent to which it is such a quality. The Scotist John of Basoles, for example, speaks of the subsequent degrees of a quality, in intensification, in terms of more perfect individuals having the same form as the individuals which precede it.[42] This line of reasoning depends on a rather important conceptual distinction: a quality has intensity not *per se* but *qua* something that exists in an individual subject. Intensification and remission belong to qualities not essentially but accidentally, as the quality exists in a given subject or *supposito*.

It is this Scotist account of intensification and remission which is taken up by Ockham, and subsequently by the Merton School, Heytesbury and The Calculator in particular.[43] The main physical problem in this account concerns the discrepancy between the idea that increases occur by addition, and the idea that in such an increase a quality, which is indivisible, persists in a subject. The Scotists did little to reconcile these ideas, and it is difficult to see how they could be reconciled. They did argue that the addition should be construed on the analogy of water to water, and not on the analogy of bricks to bricks, but all this analogy could indicate would be that the quality is undivided: it does not enable us to understand how the quality could be indivisible. John of Basoles, who, from Duhem's account,[44] seems to have given the most thought to the matter, equates the 'degree of a form' with 'the individual limited by that form':

> 'It is the same thing to compare a subject which has this form to a greater degree to a subject which has it to a lesser degree as it is to compare an individual which is more perfect in respect of this form to an individual which is less perfect.'[45]

This claim is consistent and indeed dependent on the Aristotelian conception of motion as a process in which a form which is initially potential is actualised. The description of such a process may be *relatively* straightforward, but to subject the process to a quantitative analysis is quite a different matter.

This latter problem introduces a more general issue. As Clavelin has noted, the mean degree theorem does not constitute a quantitative analysis of uniformly difform motion. Rather, it is a *reduction* of this kind of motion to uniform motion; it is a means of determining the order of magnitude of effects *by reference to a uniform intensity*. This is the whole point of the idea that uniformly difform motion *corresponds* to its mean degree of speed. Moreover, in their treatment of uniformly difformly difform motions, the Mertonians determine the effects of the changes incurred *by reference to the uniform motion* that they consider to have taken place in the first proportional part of the time.[46]

Of some importance here is the relation between logic (that is, *logica moderna*: supposition theory and the theory of syncategorematic terms) and physics in the work of the Merton School. These two are related in a reciprocal fashion. First, physical problems are treated by means of *logica moderna*. Secondly, physical principles are employed in the discussion of certain logical problems. Particularly important in the latter case is the problem of denomination: that is, the problem of how one denominates a subject in which the intensity of a quality varies from point to point. Attributes may be assumed by an individual in varying degrees of intensity or completeness. The problems arise in trying to determine the conditions under which an individual can be said to be the *suppositum* of a given term; in other words, the conditions under which it can be said to be qualified by the attribute which the term connotes.[47] One of the central aims of the work of the Mertonians is to provide a general account of how a subject is to be denominated under all conceivable circumstances of change and variation.

The basis for this kind of project is enunciated by one of the later Mertonians, Heytesbury, in a particularly clear and revealing fashion. He makes a sharp distinction between two modes of treating problems: *physice loquendo* and *sophistice loquendo*. The first mode of treatment relies on experience and the basic principles of Aristotelian physics. The second is much more general in scope: anything which is imaginable can be dealt with, and it can be dealt with by the introduction of any distinctions which are convenient (subject to two restrictions that we shall mention below). Heytesbury, and the Merton School in general, are primarily concerned with the treatment of problems *sophistice loquendo*: whether the cases dealt with have any physical significance is of little importance.[48]

In Heytesbury's work on kinematics, for example, problems about the different types of motion are explicitly posed *secundum imaginationem*. Heytesbury deals with motion *qua* change of quantity, quality and place — thus following Aristotle's threefold classification of *kinēsis*. The point of the exercise is to establish definitions of speed, or 'quantity of motion', in these three categories and there are two criteria by which the definitions are chosen: (1) the definition must accord with the 'common mode of speech' (*communis modus loquendi*); (2) it should be free of contradiction (this includes the requirement that it should not contradict any mathematical principles).[49] This project is carried out with a remarkable degree of ingenuity. For example, Heytesbury distinguishes three main kinds of difform motion with respect to place (ie., local motion). The first is difform motion with respect to time: this arises when the spaces traversed in equal times are unequal (as in free fall). The second kind is difform motion with respect to subject moved: this arises when different points of a body move with unequal (linear) velocities, as in the motion of a point at the circumference of a wheel with respect to a point near its centre. The third kind is difform motion with respect to subject and time, as when a wheel is rotated at an ever increasing rate. Heytesbury manages to reduce the second and third cases to the first by arguing that the motion of a body is to be denominated uniform or difform according as the motion of the point most rapidly moved is uniform or difform. As Wilson notes, 'this convention is justified by the fact that every magnitude, considered as a categorematic whole, moves as fast as some part of it moves, and reaches a given *terminus ad quem* as fast as this part. The principle according to which a thing is to be defined by its maximum or greatest perfection seems to be in operation here'.[50]

What is provided in Heytesbury's work on kinematics is a conceptual schema — constrained only by the principles of *logica moderna* — in terms of which any kind of variation can be described in terms of quantity and category. In dealing with the question of how one measures changes of 'quantity of motion' in difform motion, Heytesbury argues that the speed at any instant in a uniformly difform motion 'is given by the line which the point most rapidly moved would describe if it moved uniformly for a length of time at the same speed with which it moves in a given instant'.[51] This characterisation is based on a distinction between *velocitas alterationis*, which denominates the motion of the alteration or change, and *velociter alterari*, which refers to the change in denomination of the intensity of the subject altered. Wilson has shown the central reliance of these classifications on a philosophy of naming; as he points out, 'instead of attempting an exhaustive description of possible modes of variation, Heytesbury seeks to include all cases under certain arbitrary rules of denomination, which in themselves do not suffice for a complete distinction of cases'.[52]

Closely related to this question of the treatment of physical questions in purely logical terms is the issue of experiment and measurement. The Merton account of local motion (and any other physical phenomenon for that matter) completely excludes any reference to experiment.[53] No attempt was made to apply the mean degree theorem to the case of falling bodies in the fourteenth century. The first sustained attempt to do this seems to have been by Domingo de Soto in 1555.[54] Such an application is clearly rather difficult while rest and motion are categorially distinct, insofar as the transition between rest and motion cannot be treated in the same way as the transition between successive increments of a changing motion.[55] Hence, Shapere is surely wrong when he states that the failure of the Merton School to apply their work to the problem of falling bodies was due to a 'concentration on theoretical generality'.[56] The failure is, rather, inherent in the nature of their project. What Shapere calls 'theoretical generality' is, in fact, closer to a *conceptual analysis of terms*. In such a conceptual analysis the question of applicability simply does not arise: it is out of keeping with the kind of project in which the Mertonians, in their work on kinematics, are primarily engaged.[57]

The situation in Medieval dynamics — where problems are not dealt with purely in terms of *logica moderna* — is not so straightforward as this, and it is to dynamics that we must now turn. We shall concentrate on explanatory issues and in particular on the question of evidence, and on how the domain of evidence of Medieval dynamics constrains the mode of formulation of dynamical concepts.

§3   Dynamics: The Formulation of Physical Concepts

Bradwardine's account of motion *sub specie causae motivae* resolves a fundamental discrepancy in Aristotle's account. For Aristotle, the relation between motive power (F) and resistance (R) and speed (V) is such that speed is in a simple proportion to the ratio of motive power to resistance (V = k.F/R, where k is a constant).[58] There are two important assumptions in Aristotle's account. The first is that motion only occurs when F is greater than R. The second is that speed bears a proportional dependence on the F/R ratio. Bradwardine accepts these assumptions and proceeds (1) to show that the Aristotelian formulation of the proportionality is inconsistent with the first assumption, and (2) to construct a formula which is consistent with both assumptions.[59] The first matter is easily dealt with, for if we assume that a speed of 4 units is generated by an F/R ratio of 2:1 then we must conclude that a speed of 4/2 units — 2 units — is generated by an F/R ratio of 1:1, but by the first assumption when motive power are resistance are equal there can be no speed generated. On the basis of a reasoning of this kind, Bradwardine concludes that, on Aristotle's formula, 'any *mobile* could, therefore, be moved by any mover'.[60]

Bradwardine rectifies the formula by substituting a 'geometrical' for an 'arithmetical' proportionality. This takes the form of an introduction of n-tuple proportions. For an n-fold increase in speed one must take the nth power of the F/R ratio. In mathematical terms this is an important development,[61] but physically it is quite in keeping with the idea of motion as a process occurring between termini. In terms of the development of physics it remains an essentially logical exercise in which an (admittedly important) contradiction in Aristotle's account is ironed out. No experimental work was ever done on the basis of the law; indeed, as Pederson and Pihl have pointed out, it is useless as a basis for describing mechanical experiments.[62]

The Mertonian account of the dynamic conditions of motion makes essential reference to the resistance of the medium. The role of resistance, its existence as a unitary phenomenon, its status as a *sine qua non* of motion: none of these are questioned.[63] However, these doctrines were not regarded as unproblematic in the Middle Ages, and the first and third had been questioned as early as the sixth century by Philoponus. The problems arose, in the main, from the fact that while Aristotle had enumerated the causes of free fall, he had not been at all explicit about the nature of the *motor conjunctus* which maintains motion in free fall. In the thirteenth and fourteenth centuries, in particular, we find a proliferation of theories on this latter problem. Dijksterhuis[64] lists six such theories. These are all Aristotelian inasmuch as they all *presuppose a mover for motion to be maintained*. They differ mainly in their location of the mover, this location varying from the medium, to the heavy body (*grave*) itself, to the celestial sphere, the centre of the Cosmos, and finally the whole sublunary sphere.

The question of the *motor conjunctus* seems to have been kept separate from the question of the natural possibility of a void in the fourteenth century. Buridan, for example, simply shows that the Aristotelian and Avempacean accounts of motive power and resistance are incompatible, but he concludes that neither is amenable to proof (a conclusion very much in keeping with the 'instrumentalist' assumptions of the Paris Terminists). The Aristotelian account can be expressed: $V = k.F/R$. Avempace's account can be expressed: $V = k(F-R)$. Although it is unlikely that Avempace admitted the natural possibility of a void, his account of motion allows him to hold that material resistance is not essential to motion, since in a void a body moves with a speed which is simply proportional to the motive force. Motive force here is equivalent to the individual weight (as opposed to the specific weight) of the body. What is particularly interesting in this account is that the case of motion in a void is construed as physically significant, subject to legitimate physical enquiry. Moody, in considering this account of motion, concludes that 'the "essential" and "natural", for scientific enquiry, is not limited by [Avempace and his followers] to the "common sense" data of observation, but is determined by abstract and generalised terms in which a problem is formulated'.[65]

But while it is true that the range of what is analysed is not restricted by the 'common sense data of perception', this is not as true of the kind of concepts in terms of which this analysis is made. This is an important distinction and it is one that Moody fails to make. In this respect the following (very influential) remark of Moody's is worth examining:

> 'Such abstraction is a *sine qua non* of a generalised and quantitatively expressed science of dynamics; and this abstraction was admitted and practised by both the defenders and critics of Avempace's theory, in the thirteenth and fourteenth centuries. If as Cassirer and Koyré contend, this method of science is Platonist or Alexandrian, as contrasted with the "common sense empiricism" of the Aristotelians, we shall have to concede a healthy measure of Platonist and Alexandrian character to western medieval philosophy of science.'[66]

This distinction is seriously mistaken. In the first place, it is a gross oversimplification to label Aristotelianism 'empiricist', but inasmuch as it merits this description *it is precisely insofar as it is abstractionist*. That is to say, it is insofar as the concepts of Peripatetic physics are based on an abstraction from everyday experience that these concepts could be described as empiricist. In this respect, Moody's account of Medieval physics shows it to have no marked difference from the more traditional Aristotelian procedures. It is true that Avempace's account of motion allows significance to a situation which is not encountered naturally, but there is nothing intrinsically un-Aristotelian in this. Aristotle does not allow significance to the idea of motion in a void because on his account such a situation is inconceivable: he does not argue from the supposition that a void does not occur naturally. Avempace himself, from what we can tell, does not allow the natural existence of a void, but on his account motion in a void would be possible. There is a difference here but it is not one of *procedure*. The concepts of motive power, resistance and speed with which Avempace operates are general and abstract concepts (as are Aristotle's, as a matter of fact) but the *crucial issue* is: on what basis are they formulated? The answer, as we might expect, is that they are formulated on the basis of an abstraction from everyday experience. There is nothing in the work of Avempace or his followers to suggest that motive power is not simply parasitic on ideas of muscular effort or that resistance is anything other than opposition to this motive power. The fact that motive power is apparently construed as individual weight, and the fact that resistance is considered as a unitary phenomenon — that is, it is not analysed differentially into distinct kinds of resistance effect — serve to support this reading.

The subservience of physical concepts to everyday experience is not questioned in Avempace's account — nor in that of his followers or critics. We must now examine this subservience rather more closely by looking briefly at *impetus* physics. Out interest is not in *impetus* physics as such but in the manner in which physical concepts are formulated in this physics.

One of the main problems the *impetus* theory was designed to deal with was this: given the known increase in speed of falling bodies, how can a constant cause — (absolute) weight — acting in a natural way, produce a variable effect. Aristotle's explanation of this, in terms of the variation of resistance of the medium, completely contradicts his account of violent motion. Since the doctrine of natural and violent motion was not in question, it was the former that required revision, and it is such a revision that forms the basis of *impetus* theory.

The *impetus* theory originated in the work of Philóponus on projectile motion, but it found its most developed forms in the *virtus impressa* theory of Marchia and in the *impetus* theory of the Paris terminists. We shall concern ourselves with the latter. For Buridan, *impetus* is not equivalent to motion, rather, it is the cause of motion. It is not the *primary* cause of motion however: the primary cause, in the case of free fall, is weight (*gravitas*) and, in the case of projectile motion, the motive power conferred on the body by the original mover. In effect, *impetus* is a 'transferred' cause, and it can be treated as the means by which the primary cause is realised in the maintenance of motion. Here, for example, is part of Buridan's account of projectile motion:

> 'Thus we can and ought to say that in the stone or other projectile there is impressed something which is the motive force [*virtus motiva*] of that projectile. And this is evidently better than falling

back on the statement that the air continues to move that projectile. For the air appears rather to resist. Therefore, it seems to me that it ought to be said that the motor in moving a moving body impresses [*imprimit*] in it a certain *impetus* or a certain motive force [*vis motiva*] of the moving body, [which *impetus* acts] in the direction towards which the mover was moving the moving body, either up or down, or laterally, or circularly. And by the same amount the motor moves that moving body more swiftly, by the same amount it will impress in it a stronger *impetus*. It is by that *impetus* that the stone is moved after the projector ceases to move. But that *impetus* is continually decreased [*remittitur*] by the resisting air and by the gravity of the stone, which inclines it in a direction contrary to that in which the *impetus* was naturally predisposed to move it. Thus the movement of the stone continually becomes slower, and finally that *impetus* is so diminished or corrupted that the gravity of the stone wins out over it and moves the stone down to its natural place.'[67]

This passage carries the implication that in the absence of a medium a body would continue to move with a constant speed. This is made explicit elsewhere in Buridan's work: 'This *impetus* would endure for an infinite time, if it were not diminished and corrupted by an opposed resistance or by something tending to an opposed motion'.[68] *Impetus* here is an 'enduring reality' (*res permanens*). Moody identifies it with 'a condition or *state* of being in motion,'[69] but this attempt to find a precursor of inertia in *impetus* is unwarranted. *Impetus* may be an effect of the primary cause but it is itself the *cause* of uniform motion. That this is the case is clear from the passage which Moody actually cites in evidence for *his* claim:

'It must be imagined that a heavy body acquires from its primary mover, namely from its gravity, not merely motion, but also that it acquires in itself a certain *impetus* along with that motion, WHICH HAS THE POWER OF MOVING THAT SAME BODY, along with the natural constant gravity.'[70]

What is maintained or conserved, then, is the *cause* of motion and *because* this is maintained motion itself is maintained. This is not only compatible with the idea that uniform motion requires constant force, it is a restatement of this very principle: it could not be further from a statement of the inertial principle.

One of the most developed forms of *impetus* theory is to be found in the work of Benedetti. The doctrine undergoes

various changes in the period between Buridan and Benedetti
— notably in the hands of Oresme — but these will not
concern us.[71] Benedetti argues that *impetus* is an internal
motive power, rather than an external force, and that it is
preserved exclusively in a straight line. Against Aristotle, he
argues that the speed of a falling body depends on its
distance from its starting point, and not on its proximity to
its final goal. Although this account may be formally
equivalent to Aristotle's, it enables Benedetti to isolate a
moving body, conceptually, from the rest of the universe.
The suppression of the *terminus ad quem* in favour of the
terminus a quo opens up the possibility of conceptualising
force as acting in a certain direction and not towards a
certain end. Nevertheless, *impetus* remains a motive power
or force for Benedetti, and the maintenance of motion
requires this force acting as a cause. The causal mechanism
by which uniform and accelerated motion are realised is conceived in terms of *impetus*: new acquisitions of *impetus* are
produced in proportion as the body gets further away from
its starting point.[72]

In connection with *impetus* theory, it is perhaps worth
noting that Medieval mechanics had to rely almost
exclusively on everyday experience. Precision instruments
were used in astronomy and geodesy; in mechanics only a
handful of optical instruments were available.[73] However,
even given the availability of these instruments, there use was
restricted on purely theoretical grounds. This is particularly
marked in the case of optical instruments. Although one can
trace a tradition of geometrical optics from Alhazen's work
as this was taken up in the thirteenth century by Grosseteste
and Roger Bacon,[74] through Oresme, Theodoric of Freiberg
and Della Porta, the practitioners of geometrical optics were
few and far between before Kepler's work.[75] The fourteenth,
fifteenth and sixteenth centuries are dominated by the kind
of optical questions raised by Aristotle and Avicenna —
which primarily concern the psychological activity of the
perceiving subject — and in particular by the problem of the
internal senses. It is well known that Galileo encountered
considerable resistance to his use of the telescope,[76]
particularly its use in planetary and sidereal astronomy

where telescopic observation may actually contradict the observations of our unaided vision: such a case is the change in the apparent order of magnitude of the planets, a change produced by the telescope's diminution of the irradiation effect. The use of optical instruments to confirm and refine what was already known on the basis of unaided sense perception was rarely called into question. But their independent use, as correctives, was a point of contention and what resulted was an almost exclusive reliance on everyday experience.

This reliance shows itself not only in the experimental procedures, or more usually *lack* of experimental procedures, of Medieval mechanics. It also shows itself, not surprisingly, in the *concepts* of this physics. Indeed, it is the main factor determining the way in which these concepts are formulated. Koyré, in discussing Benedetti's concept of *impetus* (and in this respect it is no different from any other concept of *impetus*) has put this point admirably:

> 'Benedetti's explanation [of how circular motion is produced by *impetus* acting in a straight line] rightly appears very confused. This should not surprise us unduly: the notion of *impetus* is in fact a very confused notion. Basically, all it does is to translate into "scientific" terms a notion which is based on everyday experience, on a given of common sense. After all, what is *impetus*, *forza*, *virtus motiva*, if not, so to speak, a condensation of muscular effort and vigour? Thus it accords very well with the "facts" — real or not — which form the experiential basis of Medieval dynamics; and in particular with the "fact" of the initial acceleration of the projectile. It even explains this fact: for is not time needed for the *impetus* to take hold of a *mobile*? Everyone knows that in order to jump over an obstacle one has to "make a take off"; that a chariot which is pushed, or pulled, starts slowly and gradually increases its speed: it too takes off and gathers momentum; just as everyone — even a child throwing a ball — knows that in order to hit the ball hard he must position himself at a certain distance from it, and not too near, in order to allow the ball to gather momentum.'[77]

Benedetti proposes to found his physics on Archimedes' statics, and to establish a 'mathematical philosophy of nature'.[78] He conceives his project in conscious and explicit opposition to the qualitative physics of many of his contemporaries. He is also one of Aristotle's severest critics; in particular, he shows that bodies of the same 'nature' fall with the same speed, and that infinite rectilinear motion does not presuppose infinite space.[79] Nevertheless, his causal account of motion is fundamentally Aristotelian, and although his use of mathematics far exceeds that of his immediate predecessor, Tartaglia, the fundamentally qualitative and intuitive nature of the concept of *impetus* effectively rules out a mathematical account of motion. The kind of explanations he is seeking in mechanics are, moreover, firmly within an explanatory structure derived from Aristotle. For Benedetti, knowledge of principles is always something which has its basis in everyday experience, and he accepts the demonstrative mode of proof with all the restrictions this confers on the use of mathematics.[80]

§4   Demonstration and Evidence

For Aristotle, as we saw in the last chapter, science is defined as the demonstrative knowledge of things known through their 'causes' (*aitia*). This requires the construction of a demonstrative syllogism, which in turn requires that the *aitia* be discovered and defined in such a way that they can serve as the middle terms of demonstrations. The discovery of *aitia* is dealt with in II,19 of the *Posterior Analytics* and we have already discussed this chapter in detail. There is, however, a rather different but complementary account to be found in I,3 and II,8 of the *Posterior Analytics*, where the distinction is made between demonstration *tou dioti* — where an observed phenomenon is demonstrated from its proximate *aition* — and demonstration *tou hoti* — where the proximate *aition* is 'demonstrated' from the observed phenomenon to which it gives rise.

In the Latin commentators, these become demonstration *propter quid* and demonstration *quia est* (or simply *quia*) respectively. The problems, as might be expected, centre around the latter: the extent to which it can be called a *demonstration* at all, and whether the combination of *propter quid* and *quia* is not circular (in that we move, inferentially, from effects to cause and then back again to the same effects). There was great confusion on these questions up to the sixteenth century,[81] mainly due to the conflation of methods of presentation and methods of investigation. The clear separation of these — into *ordo* and *methodus* respectively — by such sixteenth century writers as Zabarella, Pacius and Schekius[82] allowed attention to be focused on the question of how we come by the middle terms of demonstrative syllogisms.

Zabarella claims that there are only two methods of discovery in science. These are the compositive method (demonstration *propter quid*) and the resolutive method (demonstration *quia*).[83] The two methods are combined to yield a single unified procedure — the *regressus* — which Zabarella argues is not viciously circular. In order to understand the basis of this argument it would be helpful to clarify the kinds of distinction on which it is constructed. These are fundamentally the same as those set out by Nifo in his commentary on the *Physics* (1506):

> 'Recent writers maintain that there are four kinds of knowledge. The first kind is that of the effect through the senses, or observation; the second is the discovery [*inventio*] of the cause through the effect, which is called demonstration *of sign;* the third is the knowledge of the same cause through an examination [*negotiatio*] by the intellect, from which there first comes such an increased knowledge of the cause that it is fit to serve as the middle term of a demonstration *simpliciter*; the fourth is the knowledge of that same effect *propter quid*, through that cause known so certainly as to be the middle term.'[84]

Nifo argues that since the first kind of knowledge of an effect is different from the fourth kind there is no circularity involved. He then identifies *negotiatio* with 'composition and division', leaving us with the obscure remark that 'the intellect composes and divides until it knows the cause in the form of a middle term'.

Nifo later came to reject the idea of *negotiatio* altogether,[85] thereby leaving a gap in the *regressus* which precludes our knowing which of any set of true premisses can be said to be the necessary one expressing the proximate cause. This situation arises because Nifo wishes to separate sciences such as mathematics, which is a science *simpliciter*, from natural science which, while it is not a science *simpliciter*, is nevertheless a science *propter quid*:

> 'We must say that the science of nature is not a science *simpliciter*, like mathematics. Yet it is a science *propter quid*, because the

discovery of the cause, gained through a conjectural syllogism [*syllogismus coniecturalis*], is the reason why the effect is so.'[86]

Now, as Jardine has shown, what is *conjectural* about the conjectural syllogism is the *necessity* of the premises and conclusion, not their *truth*.[87] We can be certain (at least in principle) whether the major premiss of a demonstration *propter quid* expresses the constant conjunction of two events; what we cannot be certain about is which of the constant concomitants of an effect is the proximate cause.

In the earlier passage cited, Nifo describes demonstration *quia* as demonstration 'of sign'. This is an important indicator of the kind of epistemological principles at work here. The idea of sensible things as signs of God persists throughout the discussions of *regressus*, and it becomes particularly problematic in those writers who, unlike Nifo, wish to maintain the necessity of demonstration in the natural sciences. Perhaps the best example here is Zabarella.

Zabarella treats induction as a form of demonstration *quia*. It is induction which makes sensible causes known, but demonstration *quia* includes knowledge of all causes — sensible and otherwise. Zabarella takes *ideas* to be *signs*; indeed, as Skulsky notes, they are 'at best mere signs, mere shadows, of the direct awareness of universals forever denied [to man]'.[88] Now these ideas or images are formed as a result of sense perception. When we ask in what sense they are signs we come to the crux of Zabarella's account. Natural things are simply ideas in the mind of God; in induction we move from the simulacra of these ideas *qua* Nature to their simulacra *qua* ideas in our own mind, a process which is effected by the intervention of the Holy Ghost. Hence the recognition of a necessary pattern in the world — a recognition which constitutes the establishment of a necessary cause as the middle term of a demonstrative syllogism — is a Divine gift rather than a rational act.[89] We find ourselves here back in a position which is not too far removed from Augustine's 'spiritual illumination' and Grosseteste's 'metaphysics of light'.

If this kind of account — which must ultimately result in a reduction of physics to theology — is accepted then it is difficult to see how disagreements between physicists are to be resolved. In particular, Zabarella cannot explain the 'self-evident' mark by which the mind is supposed to distinguish those general patterns in nature whose elements are necessarily connected from those which are just universally connected. There seems to be no way of deciding this issue except by fiat. The problems involved here are those that Nifo faces up to when he argues that were we to have actually grasped a necessary cause there is no basis on which we could distinguish it from merely universal causes. The basic explanatory problem is the same as Aristotle's: we simply cannot give explanations of the required kind in principle. Now it might be argued that this is a mischaracterisation of the problem: the real difficulty is that explanations of the required kind cannot be *recognised* as such, in principle. This objection would be correct, in a sense, for there is a conceptual distinction between not being able (in principle) to give a particular kind of explanation, and not being able (in principle) to recognise when an explanation of that kind has been given. Nevertheless, this conceptual distinction has no practical effectivity since there is no practical difference between a system which requires that particular kinds of explanation be given where these kinds of explanation cannot be given in principle, and a system which requires that particular kinds of explanation be given where there are, in principle, no criteria by which we can decide whether explanations given are of the required kind.

If, as I have argued, the physical theories we have discussed are based on accounts of physical explanation which generate these fundamental problems, then it is clear that, at the explanatory level, the projects of which these physical theories are part are seriously and indeed *fundamentally* misconceived. That is to say, explanatory failure occurs at the most fundamental level. By treating explanations in terms of explanatory structure — which introduces considerations of ontology, proof and evidence — I hope to have given some idea of the main issues which must be raised if a workable account of explanation in physics is to be realised. It is clearly not enough to show that a new account of physical explanation is needed; we also need to be able to articulate the kinds of factor which have a bearing on the kind of explanations which can be given in any discourse. For example, the connection between what counts as a proof and what counts as an explanation in the physical theories that we have been discussing is such that one cannot simply attempt to construct a mathematical physics while at the same time seeking the 'principles' of nature. The concept of proof in physics which we find in Aristotle and in the Medievals is consistent with the kind of projects that are set in Aristotelian and Medieval mechanics. A mathematical physics cannot be justified in terms of the explanatory structure we have described, and it is instructive that the Humanist claims for mathematics — notably those of Ramus and Melanchthon — are claims about its propaedeutic value, not about its value in physics.[90]

In these last two chapters we have examined two versions of a particular explanatory structure in physics. I have argued that the accounts of what counts as an explanation in physics in this explanatory structure are unrealisable. What is involved is explanatory failure at the most fundamental level: physical explanations of the kind required cannot be given in principle. This means that a viable physics cannot operate in terms of this structure and a new one must be provided. We shall now examine how this new structure is developed in Galileo's work. I shall *not* argue that this new structure involves no explanatory failure, but that the explanatory failures which *do* occur are not so fundamental that the whole structure has to be abandoned: these explanatory failures can be dealt with by revisions in the new structure.

APPENDIX TO CHAPTER 5:
OCKHAM's THEORY OF SUPPOSITION*

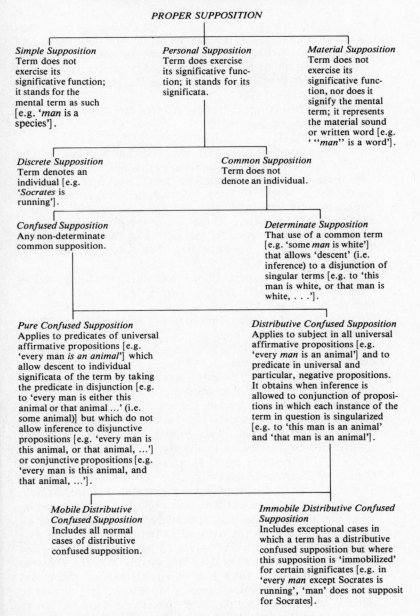

PROPER SUPPOSITION

Simple Supposition
Term does not
exercise its
significative function;
it stands for the
mental term as such
[e.g. 'man is a
species'].

Personal Supposition
Term does exercise
its significative func-
tion; it stands for its
significata.

Material Supposition
Term does not
exercise its
significative func-
tion, nor does it
signify the mental
term; it represents
the material sound
or written word [e.g.
' "man" is a word'].

Discrete Supposition
Term denotes an
individual [e.g.
'Socrates is
running'].

Common Supposition
Term does not
denote an individual.

Confused Supposition
Any non-determinate
common supposition.

Determinate Supposition
That use of a common term
[e.g. 'some man is white']
that allows 'descent' (i.e.
inference) to a disjunction of
singular terms [e.g. to 'this
man is white, or that man is
white, . . .'].

Pure Confused Supposition
Applies to predicates of universal
affirmative propositions [e.g.
'every man is an animal'] which
allow descent to individual
significata of the term by taking
the predicate in disjunction [e.g.
to 'every man is either this
animal or that animal ...' (i.e.
some animal)] but which do not
allow inference to disjunctive
propositions [e.g. 'every man is
this animal, or that animal, ...']
or conjunctive propositions [e.g.
'every man is this animal, and
that animal, ...'].

Distributive Confused Supposition
Applies to subject in all universal
affirmative propositions [e.g.
'every man is an animal'] and to
predicate in universal and
particular, negative propositions.
It obtains when inference is
allowed to conjunction of propositions in which each instance of the
term in question is singularized
[e.g. to 'this man is an animal'
and 'that man is an animal'].

**Mobile Distributive
Confused Supposition**
Includes all normal
cases of distributive
confused supposition.

**Immobile Distributive Confused
Supposition**
Includes exceptional cases in
which a term has a distributive
confused supposition but where
this supposition is 'immobilized'
for certain significates [e.g. in
'every man except Socrates is
running', 'man' does not supposit
for Socrates].

* The main presentation of Ockham's theory of supposition is to be found at the
end of Part I of his *Summa Logicae*. The schema here is condensed from the
expositions in Leff, *op. cit.*, p. 131 ff and Boehner, *op. cit.*, p.36 ff.

Notes: Chapter 5

1 As a matter of fact, this is also true of the Arab commentators on Aristotle's works, but we shall not be considering these here. For details of Muslim epistemology of the period, and particularly Aristotle's impact on it, see Rosenthal, *Knowledge Triumphant*, chs. 5 and 7.

2 Colish, *The Mirror of Language*, p.2; see also Prestige, *God in Patristic Thought*, *passim*.

3 *Cf. Meno*, 80a ff.

4 Cf. Colish, *op. cit.*, pp.5-6; also Sheldon-Williams', 'The Greek Christian Platonist Tradition', pp.425-431 and 506-517.

5 This thesis is defended in detail in Colish, *op. cit.* I have relied on Colish's work for my account of Augustinian and Anselmian epistemology; also for some aspects of Thomist epistemology. For details of the role of the *trivium* in the later Middle Ages, and especially its relation to the *quadrivium* in University teaching, see Abelson, *The Seven Liberal Arts*, *passim*.

6 Augustine, *De Magistro*, Books 2-4; cited and discussed in Colish, *op. cit.*, p.55 ff. Cf. also Kretzman, 'History of Semantics', pp.365-367.

7 Cf. Gilson, *The Christian Philosophy of St Augustine*, p.77 ff.

8 Colish, *op. cit.*, p.85.

9 In terms of 'affective status' there are two categories of hearer: believers and infidels (*sic*).

10 I am not suggesting that this transformation is peculiar to Aquinas, or even that it was initiated by him. On the contrary, there is a general decline in the study of grammar, and a corresponding concentration on the study of logic, from the twelfth century onwards. Two good general accounts of this change are to be found in Paetow, *The Arts Course at Medieval Universities*, pp.29-32 and p.39 ff; and Sandys, *A History of Classical Scholarship*, I, pp.665-678. Some attempts to revive grammatical studies were made in the later Middle Ages and indeed there was a group of scholars — the *modistae* — whose investigations into 'speculative grammar' marked an important break with the work of Donatus and Priscian. The aim of speculative grammar was the formulation of a science of grammar which concerned itself not merely with the elaboration of grammatical rules but with the reasons behind and function of grammatical rules. Moreover, the *modistae* rejected the contemporary idea that there were as many grammars as there are languages, and sought to provide, for

grammar, the unified subject matter that was a prerequisite of a science in the Aristotelian sense. The main adherent of speculative grammar was the fourteenth century scholar Siger of Courtrai, who attempted to find analogues for the various parts of speech in the Aristotelian distinctions between matter, form, substance, process and so on (cf. Wallerand, *Les Oeuvres de Siger de Courtrai*, Part I, pp.34-60). Speculative grammar was subjected to fierce attacks — which centred on the epistemological question of the relation between things, mental concepts and linguistic expressions — in the fourteenth century, most notably by Johannes Aurifaber. Nevertheless, these debates were exceptional, and it was much more usual to treat epistemological problems concerning language and thought at the logical level of the *propositio mentalis* and its *connotationes* and *suppositiones*, rather than at the grammatical level of *modus significandi* (cf. Pinborg, *Die Entwicklung der Sprachtheorie im Mittelalter*, p.137 ff and esp. p.197). Logic itself finally began to lose its central place in the fifteenth century, mainly due to the Humanist attacks on it (cf. Ong, *Ramus, Method and the Decay of Dialogue*).

11   Cf. Kretzmann, *William of Sherwood's Introduction to Logic,* and *William of Sherwood's Treatise on Syncategorematic Words*; Peter of Spain, *Summulae Logicales*. These three treatises date from the middle of the thirteenth century. For details of supposition theory see Kneale, *Development of Logic*, p.262 ff; Moody, *Truth and Consequence in Medieval Logic*, *passim*; Pinborg, *Logik und Semantik im Mittelalter*, *passim*; Boehner, *Medieval Logic*, *passim*.

12   Cf. Moody, *op. cit.*, pp.21-23.

13   Cf. Colish, *op. cit.*, p.165 ff.

14   For discussions of the details of this process see Hamlyn, *Sensation and Perception*, pp.46-51, and Gilson, *The Christian Philosophy of St Thomas Aquinas*, pp.200-235.

15   My account of these different senses is based primarily on Steneck, *The Problem of the Internal Senses in the Fourteenth Century,* ch 1.

16   Cf. in particular Aquinas, *Commentary on Boethius' 'De Trinitate'*, pp.56-62.

17   On intuitive cognition see Brampton, 'Scotus, Ockham and the Theory of Intuitive Cognition'; Day, *Intuitive Cognition*; Leff, *William of Ockham*, ch. 1; Sharp, *Franciscan Philosophy at Oxford,* p.279 ff.

18   Jardine, *Francis Bacon*, p.4.

19 For a comparison of the supposition theories of Peter of Spain (c1246) and Ockham (c1326), see Boehner, *op. cit.*, pp.19-51. The most striking technical difference between the two theories lies in Ockham's characterisation and elucidation of the various forms of personal supposition in terms of their consequences; Peter of Spain does not employ consequences in this fashion. Ockham's supposition theory is discussed at length in Leff, *op. cit.*, ch.3, esp. pp.131-139.

20 *Guilhelmi de Ockham anglici super quattuor libros* ... Lib 1, dist 2, qu 4. Cited in Moody, *Studies in Medieval Philosophy, Science and Logic*, p.145.

21 *Ibid*, p.136.

22 Ockham, *Quodlibet Septem*, VI, qu 14. Cited in Kosman, *The Aristotelian Backgrounds of Bacon's Novum Organon*, p.108. Kosman discusses Ockham's 'instrumentalism' in detail, as well as his influence on Oresme and Theodoric of Freiberg (cf. pp.101-115).

23 The essentially logical nature of Ockham's approach to physics is also discussed in Maier, *Die Vorläufer Galileis*, p.17 ff and *Zwei Grundprobleme der scholastischen Naturphilosophie*, p.74 ff.

24 The initial proportionality had been given in the Peripatetic *Mechanica*, where speed is described as being proportional to the motive power/resistance ratio.

25 Bradwardine, *Tractatus de Proportionibus*, p.125.

26 Cf. Duhem, *Le Système du Monde*, VII, pp.304-306.

27 Cf. Maier, *Zwei Grundprobleme*, pp.59-61. The idea of an 'intensive magnitude' will be discussed below.

28 Cf. *Physics*, 232a23-b20.

29 See Clavelin's excellent treatment of this is his *Natural Philosophy of Galileo*, pp.55-58. Aristotle's treatment of motion is mainly restricted to uniform motion. He is completely at sea when it comes to non-uniform motions, which he can neither define properly nor specify differentially in terms of distance traversed in given periods (his standard means of specifying uniform motions).

30 *De Caelo*, 273b30.

31 For details see Moody's essays 'Ockham and Aegidius of Rome' and 'Galileo and Avempace' in his *Studies in Medieval Philosophy, Science and Logic*, pp.161-188 and 203-286 respectively.

32 We shall restrict our discussion to the Merton treatment of local motion. For a discussion of the same kind of account applied to qualitative changes (such as condensation and

rarefaction) see Weisheipl, 'Matter in Fourteenth Century Science', pp.335-340.

33 Cf. Duhem, *op.cit.*, VII, ch.5, esp. p.480 ff.

34 Cited in Clagett, *Science of Mechanics in the Middle Ages*, pp.237 and 241.

35 Dijksterhuis, *Mechanisation of the World Picture*, p.342.

36 The same can be said for Oresme, although it cannot be denied that his graphical method constituted an important mathematical advance and enabled him to deal with elementary mechanical problems. Nevertheless, some of the more interesting applications of the method of *configurationes* are illegitimate, and were known by Oresme and his contemporaries to be illegitimate. In his geometrical treatment of uniformly difform motion, for example, he equates the area of a figure with the 'total velocity', but this is to equate an area with an infinite number of parallel lines. The literature on this topic is vast. Two of the main treatments are to be found in Maier, *An der Grenze von Scholastik and Naturwissenschaft*, pp.289-353; and in Clagett, *Nicole Oresme and the Medieval Geometry of Qualities and Motions,* pp.14-49 and p.494 ff. For discussions of later developments of the problem see Clagett, *ibid,* pp.73-111; and Drake, 'The Uniform Motion Equivalent to a Uniformly Accelerated Motion from Rest'.

37 Shapere, *Galileo*, p.59.

38 Cf. Maier, *Das Problem die Intensiven Grösse in der Scholastik*, p.9 ff; and Duhem, *op. cit.,* VII, pp.480-533.

39 *Categories*, 10b26 ff; see also Ackrill's commentary on this and subsequent passages (*Aristotle's Categories and De Interpretatione*, p.107 ff).

40 Cf. Clavelin, *op. cit.*, p.88.

41 Cf. Sylla, 'Medieval Quantifications of Qualities', p.10 ff and p.15.

42 Duhem, *op. cit.,* VII, p.507.

43 Buridan, Albert of Saxony, Peter d'Ailly and Marsilius of Inghen may also be added to the list. See Wilson, *William Heytesbury*, p.20.

44 Duhem, *op. cit.*, VII, pp.506-9; see also *ibid*, VI, pp.438-450.

45 Cited in *ibid*, VII, p.507. It is perhaps worth noting that John's discussion is primarily intended to cover changes in grace. The Medieval quantification of qualities originated in the treatment of theological problems, and particularly in Aquinas' treatment of charity in the *Summa Theologiae* (cf. Clagett, 'Richard Swineshead and Late Medieval Physics', pp.132-3). Although I have restricted my account to local motion — and

shall continue to do so — it is always important to bear in mind that the range of problems that the accounts of intensification and remission were designed to deal with were vast, and that local motion had no special importance in these accounts. Hence in challenging them much more than physics is at stake just as, in challenging Aristotle's account of *kinēsis*, much more than local motion is at stake. This is a crucial point and it cannot be ignored, as I tried to argue in ch. 1 in the discussion of the role of state-variables in the constitution of the domains of investigation of discourses. Moreover, since the Medieval account of the physical world is simply part of a more general theological project it is virtually impossible to find any special mode of treatment of 'physical' problems which is not shared by more general theological enquiries. Indeed, it is extremely difficult to identify specifically 'physical' problems in these Medieval accounts without using either Aristotelian or classical criteria.

46 Clavelin, *op. cit.*, p.85. Quantitatively, there was considerable confusion on the question of uniformly difformly difform motions. One of the main Heytesbury commentators, Gaetano di Thiene, gives as an example of a uniformly difformly difform motion the case of a body starting from rest and travelling 1 foot, 3 feet and 6 feet in equal periods of time. This would be impossible if the motion were to increase uniformly. The case is discussed in Wilson, *op. cit.*, pp.197-8.

47 Cf. Wilson, *op. cit.*, p.22 ff.

48 Cf. *ibid*, p.25 ff.

49 Cf. *Ibid*, p.115 ff.

50 *Ibid*, pp.117-8.

51 Paraphrase of Heytesbury, 'De Tribus Predicamentis', fol 39va-39vb; *ibid*, p.121.

52 *Ibid*, p.144.

53 One of the most illuminating examples in this respect is given in Bradwardine's defence of Averroes' theory of magnetism: 'There is one thing which will seem amazing to the average man [*sic*], namely, that it is just as easy to lift a magnet with a piece of iron adhering to it (whether it is underneath, on top or inside) as it is to lift the magnet alone and without raising the iron. As a matter of fact, neither does the iron resist the raising of the magnet nor does the raising of the magnet raise the iron. On the contrary, the iron moves, of itself, along with the magnet. From this it is also evident that the magnet, with or without the iron, weighs the same.' (*Tractatus*, p.123.)

54 Cf. Duhem, *Etudes sur Léonard de Vinci*, III, p.279 ff;

Clagett, *The Science of Mechanics in the Middle Ages*, pp.555-6; Maier, *Zwei Grundprobleme*, p.299; Wallace, 'The Enigma of Domingo de Soto'.

55 This particular issue is complicated by logical problems concerning the syncategorematic terms 'begins' and 'ceases'. In the *Physics* (236a3 ff), Aristotle raises the problem of the discontinuous transition between rest and movement by treating it in terms of the question of how we specify the moment in which it is first correct to say 'this has changed (or moved)'. This question receives a very extensive treatment throughout the Middle Ages. The proponents of *logica moderna* generalise the problem (as usual) to cover any imaginable case, and the terms *incipit* (begins) and *desinit* (ceases) are applied to anything that can be imagined to be at one time and not to be at another. This includes relations as well as things and motions. The standard 'modernist' treatment is to construe *incipit* and *desinit* as terms having an obscure sense which requires exposition (i.e. as 'exponible' terms): Peter of Spain, for example, considers that they implicitly contain the syncategorematic term 'immediately'. General expositions of the history of the problem are to be found in Kretzmann 'Incipit/Desinit' and in Wilson, *op. cit.*, pp.29-41

56 Shapere, *op. cit.*, p.59.

57 This raises the very interesting question of the relation between Medieval physical theory and Medieval developments in technology (where careful observations and measurements were made). This is an issue which is outside the scope of our present concern. Crombie has dealt with the problem in his 'Quantification in Medieval Physics', and I think his findings lend some support to the thesis that I have defended. Of particular interest is his conclusion that 'Medieval academic science and medieval technology were in fact two almost completely independent monologues'. (p.159) On the (unsuccessful) attempts to keep these monologues independent as late as the seventeenth century see Reif, 'The Textbook Tradition in Natural Philosophy', esp. p.23.

58 This is, at least, the standard Medieval interpretation of Aristotle, which is our primary concern. Such an interpretation is *suggested* by the discussion in *Physics*, 215b1-12, but at 250a15 ff Aristotle does not appear to treat motive power and resistance as being magnitudes of the same kind, which would preclude *any* quantitative relation being set up between them.

59 Bradwardine, *Tractatus*, esp. pp.95-105.

60 *Ibid*, p.99.

61 Cf. Boyer, *History of Mathematics*, p.288 ff.

62 Pederson and Pihl, *Early Physics and Astronomy*, p.223.

63 Bradwardine does allow the existence of a void in his *theological* writings (cf. Koyré, *Études d'Histoire de la Pensée Philosophique*, pp.79-92) but this has no bearing on the physical issues because the void is postulated to exist only before the creation of the world, when there is no motion since there is no matter. The only Latin Scholastic to argue for the natural possibility of the void seems to have been Nicholas of Autrecourt (cf. Grant, 'The Arguments of Nicholas of Autrecourt for the Existence of Interparticulate Vacua'). The Jewish fourteenth century thinker Hasdai Crescas presents a remarkably articulate and radical defence of the existence of the void, but his work had no influence on Christian thinkers. (For an account of Crescas see Jammer, *Concepts of Space*, p.76 ff.) The theoretical possibility that there might be a natural void only comes to be accepted in the Christian West in the sixteenth century (Cf. Schmitt, 'Experimental Evidence for and against a Void').

64 Dijksterhuis, *op. cit.*, p.177 ff.

65 Moody, 'Galileo and Avempace', *Studies in Medieval Philosophy*, p.250. The main followers of Avempace on the question of the relation between motive force and the void were Albertus Magnus, Aquinas, Siger de Brabant, Olivi and Duns Scotus.

66 *Ibid.* p.250.

67 Buridan, *Quaestiones super octo Physicorum libros Aristotelis*, VII, 12; cited in Clagett, *Science of Mechanics in the Middle Ages*, p.534-5.

68 Buridan, *Quaestiones in libros Metaphysicae*, XII, 9; cited in Moody, *Studies in Medieval Philosophy*, p.267.

69 Moody, *ibid*, p.268.

70 Buridan, *Quaestiones de caelo et mundo*; cited in Moody, *ibid*, p.268. Emphasis is mine.

71 Cf. Maier, *Zwei Grundprobleme der scholastischen Naturphilosophie*, Part II; Maier, *Zwischen Philosophie und Mechanik*, ch. 2; and Duhem, *Le Système du Monde*, VII-VIII.

72 CF. Koyré, *Études d'Histoire de la Pensée Scientifique*, p.160.

73 For details see the discussion of the items listed under the index entry 'Instruments, Scientific' in Crombie, *Robert Grosseteste and the Origins of Experimental Science*, p.365. See also Daumas, *Les Instruments Scientifiques*, pp.13-39.

74 On Bacon see Crombie, *ibid*, p.144 ff; Birkenmajer, *Études*

d'Histoire des Sciences en Pologna, p.284 ff; and Erickson, *The Medieval Vision*, p.42 ff. Bacon's work is highly eclectic: when it comes to the question of the physical nature of the radiation responsible for sight, for example, he attempts to run together Alhazen's doctrine of forms of light and colour, Aristotle's theory of the qualitative transformation of the medium, and Grosseteste's multiplication of 'species'. It did, however, stimulate the development of geometrical optics in the work of Pecham and Witelo — the 'perspectivists' — which resulted in a consolidation of Alhazen's work and marked an advance in the understanding of the propagation of light and its refraction in transparent substances.

75   One of the best general accounts of this development is to be found in Lindberg, *Theories of Vision from al-Kindi to Kepler*. See also Ronchi, *The Nature of Light*, for a stimulating but less reliable account. On Kepler see Straker, *Kepler's Optics*.

76   The basic Aristotelian objection to the use of lenses centres around the fact that, on the contemporary Aristotelian account of vision, it is by means of 'species' or forms that we see. These are transmitted to the eye of the observer when the object is illuminated. To interrupt the regular transmission of these species is to court disaster, since it means that we are not seeing what is really there, and the forms of species which the *nous* (or 'rational soul') abstracts from vision are no longer the forms or species of the object, but those of the image produced by the lens. Ronchi ('Complexities in the Development of the Science of Vision, p.546) has argued that the resistance to the use of lenses derives from an alleged primacy of the sense of touch, and a subordination of the sense of vision, which he claims to find in the Medieval and Renaissance periods. This curious thesis is almost certainly incorrect (cf. Lindberg and Steneck, 'The Sense of Vision and the Origins of Modern Science'). Throughout his work Ronchi has claimed that Galileo had no answer to the Aristotelian objections to the use of optical instruments, and particularly to Sizzi's *Dianoia* (1611), which is a direct attack on Galileo's use of the telescope in astronomy. He claims ('Galilée et l'Astronomie', p.169) that Galileo's own annotated copy of this work contains no technical objections, but we do in fact find two such objections (*Opere*, III, p.239 and p.244) and in these reference is made to Euclid, Witelo, Alhazen and Della Porta. There can be no doubt that Galileo did not have a good enough optical theory to support the use of lenses in celestial observation, but there can also be no doubt that Sizzi's work presented no great challenge.

Galileo himself regards the work as something of a joke (see Galileo to Salviati, 22nd April, 1611; *Opere*, XI, p.89 ff) and Kepler and Della Porta completely dismiss Sizzi's work (cf. Drake, *Galileo Studies*, Ch. 9).

77   Koyré, *Études Galiléennes*, p.50.

78   Benedetti, *Speculationem*, 167-8; cited in Drake and Drabkin, *Mechanics in Sixteenth Century Italy*, p.196.

79   For translations of Benedetti's main critiques of Aristotle see *ibid*, pp.154-164 and pp.197-223.

80   Cf. Strong, *Procedures and Metaphysics*, pp.125-134.

81   Cf. Gilbert, *Renaissance Concepts of Method*, and Randall, *The School of Padua* (Essay 1, 'The Development of Scientific Method') for details. Grosseteste seems to have been an exception in this respect; cf. Crombie, *Robert Grosseteste*, pp.61-74.

82   Cf. Jardine, *Francis Bacon*, Ch. 1. esp. pp.51-8.

83   On Zabarella see Cassirer, *Das Erkenntnisproblem in der Philosophie und Wissenschaft der Neueren Zeit*, I, pp.136-144; Randall, *op. cit.*, p.49 ff; Gilbert, *op. cit.*, pp.167-173; Jardine, 'Galileo's Road to Truth and the Demonstrative Regress', pp.296-303. On the relation between the composition/resolution distinction and the older analysis/synthesis distinction see Hintikka and Remes, *The Method of Analysis*, Ch. 9. Some of Galileo's early notes deal with the question of *regressus*, and there is evidence that these are the product of independent study on Galileo's part. See Crombie, 'Sources of Galileo's Early Philosophy'.

84   Cited in Randall, *op. cit.*, p.43.

85   In the 1543, and later, editions of the *Physics* Commentary he explicitly repudiates his earlier views on this question. Cf. Jardine, 'Galileo's Road to Truth', p.291 ff.

86   Commentary on the *Physics*; cited in Randall, *op. cit.*, p.45.

87   Jardine, 'Galileo's Road to Truth', p.293 ff.

88   Skulsky, 'Paduan Epistemology and the Doctrine of One Mind', p.354.

89   Randall, in his 'Development of Scientific Method' (*The School of Padua, op. cit.*), claims that: 'If Zabarella did not follow up the suggestion of Nifo that all natural science therefore remains conjectural and hypothetical, it was because he believed that an examination of particular instances would reveal an intelligible structure present in them; and this was precisely the faith that inspired seventeenth century science' (p.60). This statement is seriously misleading. As Skulsky has pointed out, Zabarella's schema requires God 'to sort out our

sense images according to the order of His will, so that reason might have at least a transcendent sanction for its abstractions. Indeed, much of Zabarella's confidence in the power of natural science to achieve absolute certainty seems to derive from an act of implicit faith in this *deus ex machina*' (*op. cit.*, p.356).

90 Cf. Hooykaas, *Humanisme, Science et Réforme*, p.57 ff.

CHAPTER 6

PHYSICAL EXPLANATION AS THE MATHEMATICAL FORMULATION AND RESOLUTION OF PHYSICAL PROBLEMS

§1 Introduction

GALILEO'S work does not constitute the earliest attempt to pose physical problems mathematically. In this chapter I want to establish that it does, nevertheless, provide a basis on which physical problems can be formulated mathematically. It is this basis — which we do not find in the accounts we have discussed in the last two chapters — that we shall be examining. This means that we shall be concerned not so much with the results that Galileo in fact achieved in physics, as with the kinds of results which can be produced on this basis and which could not be produced on the basis of the kinds of physics we have been discussing up to now. This 'basis' is an explanatory structure. We are interested in the explanatory structure which makes a *mathematical* physics possible. I shall argue that such an explanatory structure is to be found in parts of Galileo's later work. It may be noted that rendering a mathematical physics possible on an explanatory basis is different from actually realising this possibility. It is because of this difference that we shall not be attempting to assess Galileo's work purely in terms of its physical results. There is another kind of assessment which, in the long run, is more informative: after all, correct physical results can be purely fortuitous. The physical results which can actually be produced on the basis of any account of explanation in physics may depend on factors which are autonomous with respect to physics. In particular, they may depend on mathematical developments. This is not to deny that mathematical developments stimulate and are stimulated by developments in physics, only to say that (at least in the area that we are dealing with) mathematical developments do not *produce* developments in physics and vice versa.

The elaboration of a new explanatory structure in physics does not automatically bring with it the means whereby actual explanations may be produced in this structure. What it does do is to enable us to identify these means by specifying what function they must have and the conditions under which they can be employed.

The explanatory structure which I want to identify and assess is not one which Galileo explicitly enunciates. In Chapter 4 we were able to give an account of Aristotle's conception of 'scientific' explanation and then examine how this operated in his physics and cosmology. Our procedure in chapter 5 was not as direct, and in this chapter it will be even less so. We shall begin, in fact, by presupposing that physical problems can be posed mathematically, in principle, and ask for the conditions under which this can be done in fact. In order to be able to answer this question we must be able to recognise when a physical problem is being posed either mathematically, or at least in a form which renders it amenable to mathematical treatment. It will already be clear, from our discussion of the work of the Merton School for example, that this is not always an easy question to decide. It is a particularly difficult question in the area that we shall now be investigating. Because our familiarity with physics usually derives from a knowledge of Newtonian and post-Newtonian mechanics, we tend to identify the mathematical formulation and resolution of a problem with the presence of 'quantitative' concepts and mathematical techniques. Nevertheless, the presence of such concepts and techniques does not necessarily indicate that physical problems are being posed and resolved mathematically. The conditions under which this can occur are complex. We shall approach the issues involved here by considering two situations in which Galileo is unable to pose a physical problem mathematically

— or at least to pose it in a form which renders it amenable to mathematical treatment — despite the fact that he operates with quantitative concepts and mathematical techniques. In §'s 2 and 3 we shall examine two different kinds of attempt to render dynamical problems amenable to mathematical treatment. The requirement that the problems be treated mathematically results, in these cases, in a reduction of dynamical problems to statical and kinematical problems. In §3 we shall also examine the difference between posing a physical problem geometrically and simply *reducing* a physical problem to a geometrical one.

The kinds of difficulties inherent in these cases raise two fundamental problems. First, what is the (explanatory) *justification* for the use of mathematics in the formulation and proof of a physical theorem? Secondly, given that the use of mathematics in physics can be justified at a general level, how does one actually go about posing physical problems mathematically? The first of these problems is clearly the more basic, and it raises very general explanatory issues. It is these explanatory issues that we shall deal with from §4 onwards. The cases that we shall examine in the next two sections have been chosen in order that we might have some idea of the kind of conceptual problems that we shall have to face in §4.

§2   Problems in Formulation: Dynamics as Statics

Galileo's early mechanical works constitute an attempt to found dynamics on a basis comparable with that on which Archimedes founded statics. In the work of Archimedes, statical problems had been posed in a geometrical form, which meant that proved geometrical theorems could be introduced into the solution of statical problems, providing a standard of rigour that later sixteenth century physicists had aspired to, and which Galileo held as a paradigm of certainty in proof.[1] The priority, in terms of rigour, which Galileo conferred on geometrical proof — where premises and conclusion are explicitly distinguished, and where the conclusion is formally conclusive — is clear, for example, from this passage in the *Dialogo*:

> 'Of such are the mathematical sciences alone; that is, geometry and arithmetic, in which the Divine intellect indeed knows infinitely more propositions [than humans], since it knows all. But with regard to the few which the human intellect does understand, I believe that its knowledge equals the Divine in objective certainty, for here it succeeds in understanding necessity, beyond which there can be no greater success.'[2]

The major project of Galileo's early *De Motu* is to transform the dynamical problem of free fall into one which is amenable to geometrical formulation, and hence to rigorous solution. I want to concentrate on the kinds of physical concepts which are employed in the attempted realisation of this project.

Although Galileo does not use the word *impetus*, his concept *vis impressiva* (and in later works the more general concept *impeto*) serves much the same function as *impetus*. Whereas Buridan, for example, had accounted for the motion of projectiles in terms of corruption in *impetus*, Galileo accounts for it in terms of the diminution of the *vis impressiva*. This *vis impressiva* is imparted by the mover to the moving body. As Bonamico did before him, Galileo concludes that the *vis impressiva* gradually dies out within the moving body. In explaining this he uses the analogy of a bell,[3] invoking the *qualitas sonora* acquired by the bell on being struck. This 'sonorous quality' is contrary to the natural state of the bell — silence — and it gradually dies down after impact. Analogously, the *qualitas motiva* of a projectile is contrary to its natural state — rest — and this also gradually dies down.

Nevertheless, Galileo's treatment of *vis impressiva* is placed within a wider context than his immediate predecessors' treatment of *impetus*. He deals not only with the problem of falling bodies and projectile motion, but also with certain problems connected with hydrostatics. The domain of investigation of the *De Motu* includes the rise of light bodies in water as well as the fall of heavy bodies in air. This extension of the general problem is instructive. Archimedian statics have been concerned with the conditions under which a body in a medium will rise or fall, where these conditions are conceived in terms of departure from an equilibrium state.[4] Galileo attempts to generalise this analysis to cover the dynamical problem of the cause of differences in speeds of bodies moving through different media, and in doing so to render the dynamical problem amenable to the same kinds of geometrical treatment as the statical one. This attempted generalisation results, in the *De Motu*, in a conflation of statics and dynamics.

In that work Galileo argues, *contra* Aristotle, that there are no absolutely light bodies. All bodies are absolutely heavy but some are relatively light and move upwards because ones which are relatively heavier fall down below them. However, bodies only have their *absolute weight* in a void. Speed of fall is proportional to relative weight in a given medium, and it is only in a void that bodies fall with a speed which is that of their own weight, since only here do they have their absolute weight.

Aristotle had argued that the speed of a falling body is proportional to its absolute weight divided by the resistance of the medium. Galileo, preoccupied with the idea of reconciling the (hitherto non-geometrical) analysis of free fall with the (geometrical) analysis of hydrostatics, re-arranges the formula. He makes speed proportional to the weight of a body minus the weight of an equal volume of medium. On this formula it is clear that the speed approaches that for fall in a void as the specific gravity of the medium approaches zero. Whereas on Aristotle's formulation it is inexplicable why wood floats in water and lead does not — since they both fall in air and since the ratio of their speeds is the same in all media[5] — this phenomenon is quite compatible with Galileo's formula. Galileo equates, as mathematically identical, the buoyant effect of the medium and the artificial lightness of an impressed force as, for example, when a stone is thrown upwards. This development can be summed up in two points: (i) All physical (as opposed to geometrical[6]) bodies are *essentially heavy* and this is their only *natural* property.

(ii) Whether a body is *effectively* heavy depends on whether it is imparted with an artificial lightness which exceeds its own heaviness, and this depends on the medium and whether there is an external force impressed on the body, as in projection.

In the *De Motu*, 'force' is defined as accidental (or artificial) lightness. As such, it is measurable by the weight to which it is equal. The free fall of the body is not the result of a force because it is a natural motion. A body falls because it is (effectively) heavy. It falls to a goal — its natural place. As Westfall[7] has noted, for Galileo weight was not a force acting on a body to accelerate it, it was more like a static force which a body exercises on another that restrains it from natural motion. The idea of a *natural tendency* of falling bodies expresses Galileo's conviction that the Cosmos is a well-ordered totality. A consequence of this is the idea that 'force' is that which opposes natural motion.

So *weight* is not conceived as *force*. But since force balances weight, weight can serve as the measure of force. Rejecting the Aristotelian thesis that speed is proportional to absolute weight, Galileo puts forward the theory that speed, and hence force, is proportional to specific weight.[8] But if this is the case, how is the observed acceleration of freely falling bodies to be accounted for?

In the *De Motu* account of upward motion, when a body is thrown upwards a 'force' is impressed on it which endows the body with an artificial lightness: an effect which is the same as the buoyant effect of a fluid which alters the effective specific weight of a body immersed in it. The distinction between downward and upward motion is simply that the former, being a natural motion and hence having a definite goal, needs only an *intrinsic* cause, whereas the latter, being violent and having no aim, must always have an *extrinsic* cause. However, it is not a body's *whole* weight which carries it downwards: or, at least, not in a resisting medium. Generally, a body will rise only if the impressed force — levity — is greater than its weight. When its weight and the impressed levity — which dies down — are in equilibrium, the body comes to rest. When it subsequently begins to fall the force acting on it is measured by the surplus of its weight over its levity, since the levity retards the fall of the body. Because of this, Galileo can accommodate the fact that falling bodies accelerate to the thesis that they have a constant speed which is proportional to their specific weight, simply by maintaining that if dropped from high enough the speed will become uniform. That is to say, the motion becomes uniform when the impressed levity completely disappears and only weight acts.[9]

Galileo's attempted dynamical analysis of motion in the *De Motu* remains based on considerations of weight alone: the link between weight and speed is not severed. His use of hydrostatics enables him to reject the relation between absolute weight and speed, but no new dynamical concepts are formulated. All that happens, in fact, is that concepts from statics are made to do the work of dynamical ones. This conflation of statics and dynamics prevades the whole of Galileo's work right up to the *Discorsi*. There, *impeto* — or impelling force — is measured in terms of static force:

> 'It is manifest that the impetus [*impeto*] of a heavy body is as great as the minimum resistance or force [*resistenza o forza minima*] that suffices to fix it and hold it [at rest]. I shall use the heaviness of another moveable for that force and resistance, and [as] a measure thereof.'[10]

The concept *impeto*, while it is in many respects a simple development of *impetus*, is also closely connected with the force which is needed by a simple machine to move a heavy body against its nature. *Impeto* is not directed towards horizontal motion but towards the vertical motion by which it is measured. It does not refer to an extrinsic cause which alters the body's state of rest or motion; rather, it refers to the capacity of a body in motion to raise itself to the height from which it fell in acquiring that motion.[11]

In discussing the concept of *impetus* in the last chapter we noted its essentially qualitative and intuitive nature. The standard translation of *vis impressiva* and *impeto* as 'impetus' is instructive inasmuch as it brings out the close relation between these two Galilean concepts and the earlier one. It also indicates the problems inherent in the attempt to transform these concepts by treating them in terms of a quantitative statics.

§3 Problems in Formulation:
Dynamics as Kinematics

In the discussion of an objection to the earth's diurnal motion in the 'Second Day' of the *Dialogo*,[12] another situation arises where Galileo is unable to transform a problem and its solution mathematically, because of inadequate or inappropriate physical concepts. Here a rather more concerted attempt is made to pose and resolve a problem mathematically. In this attempt, however, Galileo conceptualises the problem on a purely kinematic basis, which is quite inappropriate given the nature of the issues which have to be dealt with. His solution is consistent with many of his dynamical suppositions, but it has absurd consequences.

In his discussion of the effects of the earth's diurnal motion in the 'Second Day' of the *Dialogo*, Galileo deals

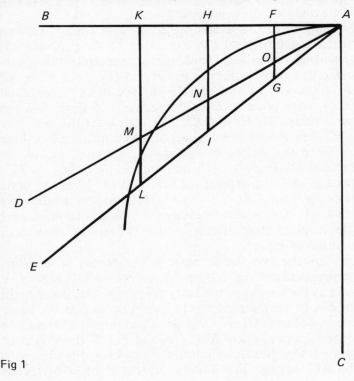

Fig 1

with four objections. They are that, if the earth revolved about its axis, then (1) bodies would not fall perpendicularly;[13] (2) bodies which are projected — such as cannon balls — would have a different range if projected in the direction of the earth's rotation than if projected in the contrary direction;[14] (3) bodies which do not adhere to the earth — such as birds and clouds — would not be able to keep up with the earth's rotation;[15] (4) bodies would be thrown outwards. We shall concern ourselves with the fourth objection. It begins with a statement of the fact that rapidly rotating wheels, such as treadmills and the potter's wheel, cause objects on their surfaces to be thrown off at a tangent. Sagredo argues:

> 'For that reason it has appeared to many, including Ptolemy, that if the earth turned upon itself with a great speed, rocks and animals would necessarily be thrown towards the stars, and buildings could not be attached to their foundations with cememt so strong that they too would not suffer similar ruin.'[16]

The reply to this objection is remarkably cogent. Galileo constructs a geometrical representation of the problem (cf. Fig 1) in which the uniform tendency to tangential motion is represented by equal distances AF, FH and HK along a line AB which is tangent to the earth's surface at A. The equal distances correspond to equal increments of time, and the perpendiculars FG, HI and KL — which meet the arbitrary line AE — represent the degrees of speed acquired in these times. Since the degree of speed KL, acquired in AK, is represented relatively to HI, acquired in AH, and to FG, acquired in AF, the degrees KL, HI and FG have the same ratios as the times KA, HA and FA. As these perpendiculars tend towards A — which represents the 'first instant of time and the original state of rest' — the degrees of speed will decrease *ad infinitum*.

Correlatively with the decrease in the degree of speed as the perpendicular from AB to AE approaches A, there is a decrease as the weight of the body decreases. The decrease in weight can be represented by the line AD, so that the angle DAB is less than EAB. For the same increments of time, the degrees of speed are now KM, HN and FO. If the weight of the body diminishes indefinitely the downward tendency to motion will increase indefinitely, so that the line of speeds

AGIL will tend toward the horizontal line *AB*, for the angle *BÂL* tends indefinitely towards zero. However, this seems to confirm Simplicio's objection that light objects would be thrown off the surface, since there appears to be a combined diminution *ad infinitum* in the downward tendency of light objects. Salviati answers this by referring to the 'horn' angle — the angle formed between the tangent and the circumference. The argument can be represented most simply as follows:[17]

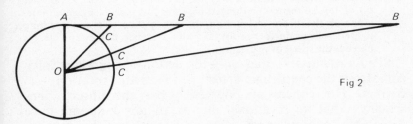

Fig 2

As the tangent along which the body would be projected approximates more and more closely to the circle in which it moves, the ratio between the length of the tangent and that of the secant increases. That is to say, as the angle $A\widehat{O}B$ diminishes, *BC* diminishes in proportion to *AB*. Thus the force acting along *AB* would need to be much greater than that acting along *AC* if it were to overcome this latter force.

Boyer[18] has claimed that Galileo's argument — if presented in a more rigorous fashion — is equivalent to saying that the function sec*x*-1 is an infinitesmal of a higher order with respect to the angle *x*. This is true, but such a generalisation would have been alien to Galileo's wholly geometrical presentation, where it is ratios rather than functions which are at stake. Further, Galileo is not even consistent in his interpretation of the components of his geometrical representations. His treatment of *Fig* 1, for example, is confused in that its abscissae and ordinates originally represent time and speed[19] and are then construed in terms of spaces of fall and horizontal motion.[20]

The 'horn' angle argument has the absurd consequence that even if a body's speed along the tangent were a million times faster than its speed along the secant it would not be extruded from the earth. When Simplicio puts this forward as an objection he receives the following curious reply:

> 'Saying this, you say what is false; not from any deficiency in logic or physics or metaphysics, but merely in geometry. For if you were aware of only its first principles, you would know that a straight line may be drawn from the centre of a circle to a tangent [cf. *Fig* 2], cutting this in such a way that the portion of the tangent lying between the contact and the secant will be a million or two or three million times greater than that portion of the secant which remains between the tangent and the circumference; and by degrees as the secant approaches the contact, this proportion becomes greater *ad infinitum*. So there is no danger, however fast the whirling and however slow the downward motion, that the feather (or even something lighter) will begin to rise up. For the tendency downward always exceeds the speed of projection.'[21]

Galileo's premiss — that only the action of a greater force can offset the centrifugal effects of the earth's rotation — is correct. His conclusion, however, has the absurd consequence that no rotational motion of any kind can have detectable consequences. The difficulty arises because instead of examining centrifugal force, Galileo simply treats it in terms of the distance which it tends to remove bodies from the earth's surface (which is particularly interesting in the light of the fact that for *Aristotle*, as we noted above, in measuring the magnitude of a motion in terms of distance traversed one is measuring *the same thing* as when one determines the 'motive power' responsible for the motion), and on his geometrical representation this distance is almost undetectable near the point of separation. Further, when it comes to the question of comparing bodies of different weights the geometry completely takes over and two distances are simply compared.

What has happened here is that instead of the physical problem being posed geometrically it has simply been reduced to geometry. This kind of reduction is prevalent in the early attempts to pose physical problems mathematically. In Galileo's first (1604) attempts to treat free fall geometrically, for example, a triangle is constructed the perpendicular of which represents distance fallen.[22] Thus speeds are plotted against distances, and not against times. This geometrical representation is simply an abstraction from everyday experience. It is simply a translation of what is seen into a geometrical form. The problem is not posed geometrically; rather, some height from the earth's surface is represented as a vertical line and the other parameters are constructed around this line, thus leaving no logical space for the time parameter. The problem is reduced essentially to a drawing exercise. In 1619 Descartes (typically) does exactly the same thing.[23]

It is Galileo's inability to pose the problem of centrifugal force in a geometrical form — as opposed to reducing it to a geometrical problem — which generates his absurd conclusion. What is particularly curious about the conclusion, however, is the fact that Galileo himself seems to recognise that it is unsatisfactory — as is clear from his treatment of rotating wheels later in the 'Second Day' of the *Dialogo*. In his initial treatment of the 'horn' angle (cf. Fig 2), the geometrical treatment makes centrifugal force independent of the speed of rotation, and it makes the dispersion effect of centrifugal force independent of the radius of the circle. When he returns to the problem, however, these difficulties are replaced by quite new ones, which arise from his attempt to use an analogy with statics, and particularly with the principle of virtual motions.

In the second treatment of centrifugal force,[24] it is established that, for two bodies near the rim of a wheel, 'two equal movable bodies will equally resist being moved if they are made to move with equal speed. But if one is to move faster than the other, it will make the greater resistance according as the greater speed is to be conferred upon it'.[25] That is, the magnitude of the centrifugal force is proportional to the angular velocity of the wheel. If, however, angular velocity is constant, what effect does the diameter of the wheel have on the linear velocity of a body near its rim? Galileo treats this problem in terms of principles taken from statics. The case of two bodies of equal weights with the same angular velocity is taken as being analogous to the case of two equal weights in equilibrium on a balance. The second case is then taken as being equivalent to two unequal weights in equilibrium on a steelyard. The geometrical problem is then set up as follows. 'Let there be two unequal wheels around [the] centre *A*,*BG* being on the circumference of the smaller and *CEH* on that

of the larger, the radius *ABC* being vertical to the horizon.
Through the points *B* and *C* we shall draw the tangent lines
BF and *CD* and in the arcs *BG* and *CE* we take two arcs of
equal length, *BG* and *CE*. The two wheels are to be understood as rotating about their centre with equal speeds in such
a way that two moving bodies will be carried along the
circumferences *BG* and *CE* with equal speeds. Let the bodies
be, for instance, two stones placed at *B* and *C*, so that in the
same time during which stone *B* travels over the arc *BG*,
stone *C* will pass the arc *CE*.'[26]

Galileo's subsequent argument can be summarised as
follows. *B* is projected more forcefully than *C*. The reason
for this is that in order to offset motion along the tangent *BF*
a body's weight would have to be withdrawn as far as *GG'*,
whereas in order to offset motion along *CD* the body would
have to be withdrawn to *EE''*. Since *EE'* is less than *GG'*, and
since the difference is greater the difference in sizes of the
wheels, the force required to prevent projection will have to
compensate for different withdrawals.[27] Applying the
principle of virtual velocities, since *GG'* and *EE'* correspond
to motions made in equal times, the small wheel will impress
on *B* a greater virtual velocity along the tangent than the

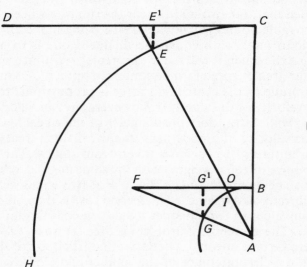

Fig 3

larger wheel impresses on *C*. Multiplying these velocities by the identical weights of the bodies, it follows that the small wheel impresses a greater *impeto* on *B* than the larger wheel impresses on *C*. That is, as the radius is greater so the centrifugal effect is smaller. On a rotating body the size of the earth, the effect would be undetectable.[28]

The confusion involved in this reasoning is as serious as it is blatant. Galileo assumes that bodies of the same weight traverse the equal distances *CE* and *BG* in equal times. But since the circumference on which *BG* lies is less than the circumference on which *CE* lies, to confer equal *linear* velocities on *B* and *C* is to confer a *lower angular* velocity on *C*. If the angular velocities are the same then it is clear that *C* will reach *E* in the same time that *B* will reach *I*. Since *CE* is greater than *BI*, *C* will be flung further than *B* if they both leave their respective wheels after the same period of time.

The problem which Galileo poses is simply irresoluble without reference to centripetal force, since this alone can prevent the appearance of centrifugal effects. Galileo himself seems much more concerned with demonstrating that circular motion has no centrifugal effects than with examining centrifugal force itself.[29] The reasons for this are interesting. Throughout Galileo's work the world is still conceived as a finite ordered whole. The geometrical space of Euclid is infinite, of course, and it is this space that geometrical demonstrations operate with. Nevertheless, this geometrical space is contrasted with the 'physical' space in which 'real events' occur. The rigid separation of the two spaces, which is central to the *Dialogo*,[30] demonstrates how little Galileo had been able to integrate his mathematical and experimental procedures at this point. Physical space had originally been conceived as that which is enclosed between the motionless earth and the sphere enveloping the Cosmos. Galileo's conception of the world as a sphere of finite radius was in no way challenged by the heliocentric theory since, in the absence of an annual parallax of the fixed stars, the size of the earth's orbit must be neglected in relation to the radius of the sphere of the fixed stars.[31] A world of finite size precludes the existence of rectilinear motion with a uniform velocity.

As well as being finite, Galileo's world is also ordered. That 'the parts of the world are disposed in a perfect order' is a premiss of his argument against uniform rectilinear motion continuing indefinitely. Because uniform rectilinear motion would require infinite space for its maintenance, a body in such a motion would 'move towards a place where it is impossible to arrive'. Again, 'one may reasonably conclude that for the maintenance of perfect order among the parts of the universe, it is necessary to say that movable bodies are movable only circularly'.[32] However, 'circular motion is never acquired naturally without straight motion to precede it; but, being once acquired, it will continue perpetually with uniform motion'.[33] This circular motion which continues 'perpetually' cannot, however, be construed as an inertial motion, as Koyré and others have argued, for as Galileo himself says:

> "I did not say that the earth has neither an external nor an internal principle of moving circularly; I say that I do not know which of the two it has. My not knowing this does not have the power to remove it'.[34]

The two examples that we have considered in this section are serious anomalies in the circular inertia account, for if the circular motion of terrestrial bodies (or, more strictly speaking, 'earthy' bodies — *cose terrestri*[35]) is inertial it becomes wholly obscure why Galileo should devote so much energy to attempting to prove that objects are not flung off the earth. If they were subject to circular inertia there would be no reason why they should be projected at a tangent to the earth's surface. Further, the thesis of circular inertia amounts to the claim that angular velocity is preserved. This angular velocity cannot be preserved with respect to the centre of the earth — since this would rule out the 'natural' circular motion of the planets — or with respect to the centre of the sun — since this would rule out the earth's 'natural' diurnal motion. The problems become even more severe when we consider that the 'natural' motion of the earth includes two different kinds of circularity, in that it undergoes both annual and diurnal motion.

Coffa has argued that Galileo is in fact seeking to elucidate the behaviour of bodies in a force-free system.[36] In order to be able to do this he has to neutralise the effects of gravitation. In the case of a parallel force field, this neutralisation is effected by positing a plane surface on which bodies move. This relies on the 'Archimedian' assumption that, because of the earth's size, areas on its surface can be treated as lying on the surface of an infinitely large sphere; that is, they can be treated as true geometrical planes. Since the terrestrial force field is central and not parallel, however, Galileo goes on to eliminate the Archimedian assumption and (erroneously) infers that bodies will preserve their velocities on spheres concentric with the earth. In both these cases, the bodies are posited as lying on a material surface, the effect of which is to neutralise gravitation. If this material surface is limited, however, the body falls over its edge. The resulting motion consists of a uniformly accelerated motion downwards together with a rectilinear and uniform motion. Coffa's claim is that this latter motion is the one which would have obtained, in the absence of gravity, if the plane had not been there.

This interpretation is problematic, for it implicitly ascribes to Galileo a workable concept of force. This is not really justified. Galileo treats gravity, for example, as being simply equivalent to force to move towards a common centre. In extrapolating Galileo's arguments to force-free systems Coffa arrives at the possibility of a body undergoing perpetual uniform rectilinear motion in an otherwise empty universe. But if gravity — construed in terms of a common centre towards which bodies fall — is removed, so is the determinate nature of the motion which a body undergoes. Extrapolated to cases which no gravity acts, bodies would simply continue to move 'in the same way', by the principle of sufficient reason. This means that in the case of a projectile launched with a 'violent' motion — and therefore with an initial rectilinear component — the body would move along a rectilinear path. But consider the case of a stone resting on a tower. If the stone were left undisturbed and the rest of the universe removed the stone would continue to move along the same *circular* path.[37] Galileo never indicates that there is a physically significant change of direction in circular motion and this, after all, is the really crucial point.

Galileo's treatment of bodies on rotating surfaces is undeniably sophisticated, but because of his inability to deal with forces he cannot conceptualise the problem properly and hence cannot pose it in a mathematically viable form. In this section and in the last one, I have tried to show that his attempts to deal with dynamical problems are generally based on an illegitimate extension of areas such as kinematics and statics which he can handle geometrically. The treatment of dynamical problems in their own right is extremely difficult — even at an elementary level — without the concept of a sum of infinitesimals (i.e. integration). Lack of the appropriate mathematical concepts is one of the crucial factors which blocks the development of Galilean dynamics. What results is a displacement of dynamical problems into kinematics and statics, where problems can be posed — in the main — in a geometrical fashion.

§4 The Mathematical Transformation of Physical Problems

Up to now in this chapter, we have discussed two kinds of case in which attempts to pose problems mathematically have failed. This leads us to ask what conditions must be fulfilled if a physical problem is to be transformed, and solved, mathematically; and this, in turn, requires a return to our fundamental problem: What is the justification for the use of mathematics in the formulation and proof of a physical theorem? We have seen that a physical problem cannot be solved mathematically unless the appropriate physical concepts are available. In dealing with the question of justification, we must also deal with the question of the conditions under which the appropriate physical concepts can be formulated.

There are two questions which must be answered initially: why is the problem of justification fundamental? and, what kind of problem is it? We are going to answer these questions by an extended examination of the conditions under which the successful formulation and proof of a particular theorem is possible. This theorem is the First Theorem on the motion of projectiles in the 'Fourth Day' of the *Discorsi*. What we shall have to say will be applicable to much more than this Theorem, as will soon become evident, but it is useful to have one example to focus on. The Theorem states that: 'The line described by a heavy moveable, when it descends with a motion compounded from equable horizontal and natural falling [motion], is a semiparabola.'[38] The proof of the Theorem is straightforward, and we can simply quote Galileo's demonstration:

'Imagine a horizontal line or plane *AB* situated on high, upon which the moveable is carried from *A* to *B* in equable motion, but at *B* lacks support from the plane, whereupon there supervenes in the same moveable, from its own heaviness, a natural motion downward along the vertical *BN*. Beyond the plane *AB* imagine a line *BE*, lying straight on, as if it were the flow or measure of time, on which there are noted any equal parts of time *BC*, *CD*, *DE*; and from points *B*, *C*, *D* and *E* imagine lines drawn parallel to the vertical *BN*. In the first of these, take some part *CI*; in the next, its quadruple *DF*; then its nontuple *EH*, and so on for the rest according to the rule of squares of *CB*, *DB*, and *EB*; or, let us say, in the duplicate ratio of those lines.

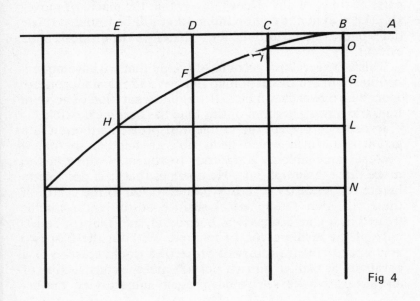

Fig 4

'If now to the moveable in equable movement beyond *B* toward *C*, we imagine to be added a motion of vertical descent according to the quantity *CI*, the moveable will be found after time *BC* to be situated at point *I*. Proceeding onwards, after time *DB* (that is, double *BC*), the distance of descent will be quadruple the first distance, *CI*; for it was demonstrated in the earlier treatise [Theorem II on Naturally Accelerated Motion] that the spaces run through by the heavy things in naturally accelerated motion are in the squared ratio of the times. And likewise the next space, *EH*, run through in time *BE*, will be as nine times [*CI*]; so that it manifestly appears that spaces *EH*, *DF*, and *CI* are to one another as the squares of lines *EB*, *DB*, and *CB*. Now, from points *I*, *F*, and *H*, draw straight lines *IO*, *FG*, and *HL* parallel to *EB*; line by line, *HL*, *FG*, and *IO* will be equal to *EB*, *DB*, and *CB* respectively, and *BO*, *BG*, and *BL* will be equal to *CI*, *DF*, and *EH*. And the square of *HL* will be to the square of *FG* as line *LB* is to *BG*, while the square of *FG* [will be] to the square of *IO* as *GB* is to *BO*; therefore points *I*, *F*, and *H* lie in one and the same parabolic line.

'And it is similarly demonstrated, assuming any equal parts of time, of any size whatever, that the places of moveables carried in like compound motion will be found at those times in the same parabolic line. Therefore the proposition is evident.'[39]

This is a successful account of a kind that we have not encountered in our discussion up to now, and the main concern from here onwards will be to determine what kind of account it is, and how an account of this kind has become possible.

In order to present the problem of projectile motion in a geometrical form which he is able to handle, so that it becomes amenable to a rigorous treatment, Galileo has to make three 'assumptions'. He assumes that it is possible to ignore the tendency of heavy bodies to fall to the centre of the earth, that the spherical surface of the earth can be treated as a true geometrical horizontal, and that the resistance of the medium can be ignored. Without the first two assumptions the trajectory of projectiles would cease to be a parabola, as Galileo himself notes,[40] and the construction of an alternative curve would present unnecessary mathematical difficulties. Without the third 'assumption' we could not suppose the motion of bodies to be uniformly accelerated.

Now the medium obviously has *some* effects on the motion of a body, and in the *Discorsi* Simplicio presents a formidable objection to the third assumption:

> 'In my opinion it is impossible to remove the impediment of the medium so that this will not destroy the equability of the transverse motion and the rule of acceleration for falling heavy things.'

That is to say, the third assumption departs from conditions not realisable in practice. Galileo's whole procedure in physics is in question here. Is not his theory only applicable to an 'ideal' situation whose relation to 'reality' it is impossible to estimate? As Simplicio says:

> 'All these difficulties make it highly improbable that anything demonstrated from such fickle assumptions can ever be verified in actual experiment.'[41]

An objection very similar to this had been raised in the *Dialogo*, again by Simplicio:

> 'I would not do Plato such an injustice [*viz*, to say that the study of mathematics disturbed his reason], although I would agree with Aristotle that he plunged into geometry too deeply and became fascinated by it. After all, these mathematical subtleties do very well in the abstract, but they do not work out when applied to sensible and physical matters.'[42]

This kind of argument raises very general issues. All Galileo's achievements are at stake here, and this is why the question of the justification for the use of mathematics is fundamental. If he has simply been dealing with 'idealisations' his mathematical proofs are fine, but they can tell us nothing about 'reality' — that is, physical and sensible matters. Such an objection to mathematical physics was commonplace in the late sixteenth and early seventeenth centuries. Guidobaldo — a patron of Galileo's — had considered that the only solution to the problem was to shun any 'simplifications' and to attempt to give a full mathematical account of sensible reality. This ill-conceived project is unrealisable not just because of the mathematical complexities it would involve but, much more importantly, because it could not produce a general mathematical physics: it could not provide a basis for distinguishing the general parameters in terms of which general physical laws are formulated from the many other variables which become operative when one deals with specific cases. Moreover, it concedes far too much to the Aristotelian equation of 'reality' with whatever is sensible.[43] I shall return to these two points in detail later. For the moment, it is more important that we are clear about the precise nature of the Aristotelian objection to a mathematical physics.

We have noted, in the last two chapters, that the kernel of the Aristotelian objection to the use of mathematical proofs in physical theory is that mathematics is simply not applicable to reality. Mathematics is applicable to particular kinds of abstractions from reality which are different in kind from reality itself: hence, 'geometrical exactitude should not be sought in physical proofs'.[44]

The Aristotelian distinction between the domains of mathematics and physics is, as we have seen, not simply an *ad hoc* feature of his account. Furthermore, the Aristotelian canon of explanation — the theory of the demonstrative syllogism — requires that this distinction be conceived in such a way as to preclude the mathematical transformation of physical problems. Since each science is defined in terms of its subject genus — physics deals with those things which have an independent existence but change, mathematics with those things which have no independent existence but do not change — and since in the demonstrative syllogism what are generated are the essential properties of the subject genus, the principles of one science cannot be used in the demonstrations of another. In the demonstrative syllogism we are concerned with attributes which inhere *essentially* in a subject. Where subject genera are distinct, distinct principles or *archai* are needed. The essences of physical phenomena are quite different from the essences of mathematical phenomena, and since what we are trying to do in science (*epistēmē*) is to give an account of essences, mathematical proofs are simply inappropriate in physics.

We have also noted that the intuitive and experiential nature of the concepts formulated in Peripatetic mechanics generally is no accident: it is intimately connected with

Aristotelian epistemology, and is reflected in the account that we are given of commensurate universals in Aristotle's work. The simple and intuitive abstraction from everyday experience which forms the basis of *impetus* physics, for example, is a procedure which is not merely encouraged by Aristotle's own stress on sense perception; it is the only procedure compatible with the requirements concerning evidence in the explanatory structure within which Aristotelian and *impetus* physics work.

I have noted that this explanatory structure operates with a notion of proof which precludes the use of mathematics in physical explanations. The introduction of mathematical formulations and proofs into physics requires the introduction of a concept of proof which is incompatible with that of the Aristotelian explanatory structure. The idea that explanations must take the form of a demonstrative syllogism in which universal and essential relations are exhibited must also be rejected. An explanation now becomes the transformation of a domain of investigation into a system of kinds of entity bearing definite and mathematically determinate relations to one another, together with a system of concepts which not only specify evidential situations in unambiguous terms, but also specify access to these evidential situations.

In the following sections, I propose to examine what is involved in this new account of explanation in physics. First, however, there are some considerations which may be apposite by way of an introduction. Examinations of Galileo's work in the *Discorsi* have usually centred around his introduction of mathematical proofs into kinematics. Until the work of Cassirer, Koyré and Burtt, the basis for this introduction was never really examined. The merit of these authors — and Koyré in particular — was to raise the conceptual issue of the justification for this introduction. Their answers to this question varied in details, but the common consensus was that mathematics was introduced on a Platonist — or even Pythagorean — basis (although Cassirer's position on this is not always clear). The claim was that arithmetical and geometrical entities actually figured as the *ontology* of physics, the implication being that Galileo conceived physics as merely a branch of mathematics. Nevertheless, while some of Galileo's statements — such as his remarks about the 'book of nature' in *Il Saggiatore* — could be construed as being of Platonist or neo-Platonist origin, it has never been made quite clear how these statements bear on what Galileo actually does in the *Discorsi*. Galileo makes many incompatible statements about what he is doing, and in at least one place claims that it is he, and not his opponents, who is the true Aristotelian.[45] It is clear, then, that reference to what *Galileo* thought he was — Platonist or Aristotelian — is not going to get us very far. Further, the terms 'Platonist' and 'Aristotelian' are subject to wide variations of meaning,[46] and in this context they are not always informative.

The account of explanatory structures which was put forward in Part I, and which has been used to examine Aristotelian and Medieval physics in the last two chapters, is designed to provide us with a more rigorous and informative way of isolating and examining the factors which have a role to play in physical explanation. Since we are dealing with explanation alone, many questions which are raised in more general histories of physics are not relevant in the present context. This is particularly true of the problem of experiment. The claims that Galileo was an *a priorist*, for example, are often countered by claims about his ability as an experimenter. Such a debate has little bearing on the theoretical problem of how we construe the role of experiment in Galilean mechanics, and it is precisely this role that we are interested in. I want to show, in fact, that the role which is conferred on experiment in the *Discorsi* is of absolutely central importance. It involves a complete reconstitution of the domain of evidence in physics, and a complete reconsideration of the means whereby concepts are formulated; indeed, I shall argue that it constitutes the *sine qua non* of the mathematical formulation of physical problems.

§5 Experiment

The role of experimentation in Galileo's work is notoriously difficult to assess. First, there are seemingly derogatory remarks about experiments throughout his writings. In the *Discorsi*, for example, we are told that:

> 'The knowledge of one single effect acquired through its causes opens the mind to the understanding and certainty of other effects without need of recourse to experiments.'[47]

Secondly, Galileo sometimes describes experiments that he has not, and in some cases could not possibly have, performed.[48] Thirdly, on the few occasions on which he does present experimental results these are often wildly inaccurate. For example, with regard to the attempted determination for a value for the constant of acceleration,[49] Koyré remarks:

> 'What an accumulation of errors and inexactitude! It is obvious that the Galilean experiments are completely worthless. The very perfection of their results is a rigorous proof of their incorrection.'[50]

There is some justification for this outburst. Although Galileo realises that he cannot perform direct measurements and, with great ingenuity, substitutes motion on an inclined plane, the operation is performed by rolling a bronze ball down a 'smooth and polished wooden groove'. He assumes, incorrectly, that a ball rolling down a plane is equivalent to its gliding (without friction) down the same plane. The device in which he measures time — a vessel with a small hole in the bottom through which water runs, to be collected and weighed afterwards — is, Koyré claims, hopelessly inaccurate and had been superseded over a millenium earlier by the Roman water clock. The result, determined with 'such precision that, as I have said, these operations repeated time and time again never differed by any notable amount',[51] gave less than half the correct value!

This antipathy to, and lack of ability in the manipulation of, experiments is taken by Koyré to be a consequence of Galileo's Platonism. The interpretation is reinforced by the dialogue form of many of Galileo's works, together with his staunch defences of Plato and attacks on 'his Peripatetic majesty' throughout the *Dialogo* and the *Discorsi*. For Koyré:

> '[Galileo] was obliged to drop sense perception as the source of knowledge and to proclaim that intellectual, and even *a priori* knowledge, is our sole and only means of apprehending the essence of the real.'[52]

If this is to credit Galileo with being a Platonist, the case is further strengthened by the role that 'simplicity' plays in his work. The appeal to simplicity is not merely an expedient in Galileo. We have noted in §3 that the Galilean world is an ordered totality. This order in the world is represented in its simplicity.[53] Of two explanations which are supported equally well, the simpler must be superior: the greater simplicity of a theory is a proof of its closer agreement with 'reality'. Seen in these epistemological terms, 'reality' does seem to become a function of 'reason' for Galileo, which would indicate a strong Platonist base in his work. What is problematic about this kind of account, however, is whether the terms in which the question is posed can provide a suitable basis for an examination of the role of experiment and mathematics — and the relation between these two — in Galileo's work.

It may be worthwhile, at this point, to summarize briefly the terms of the Plato/Aristotle debate on the relation between 'reason' and 'reality', and consequently on the nature and role of mathematics. On the Aristotelian theory, mathematics is not applicable in physical proof or explanation because what is real (in the sense of being prior in the order of knowing) is what is given to the senses, and mathematics is only applicable to certain kinds of abstraction from reality. This is not a devaluation of mathematics; rather, it is a claim that mathematics must be restricted to its own domain. On the Platonic, or more strictly speaking neo-Platonic, theory, the realm of numbers, conceived as efficient causes, is not only a realm in its own ontological right, it is the realm of 'reality' which is the cause of the world of our sense perceptions.[54] We have seen the self-defeating consequences of the Aristotelian 'realist' or 'essentialist' approach to questions in physics. But, we may well ask, is the neo-Platonic or 'idealist' approach any better? Is the denigration of mathematics any better than the denigration of experiment in physics? The answer is that it is not. In fact, what is instructive about the two theses is not so much the contrast between them as the presuppositions which they share. On both theses we have two realms, so to speak: the realm of sensible bodies and their sensible properties and relations, and the realm of reason and mathematical truths.

Now it is clear that Galileo does accept the terms of the Aristotle/Plato debate — and that he comes down on the side of Plato — in parts of his work. This is reflected in his epistemological distinction between primary and secondary qualities and the ontology which results from this distinction, an ontology in which the realm of reality is that of primary qualities.

In the *Il Saggiatore* of 1623 Galileo enumerates what he considers to be the primary qualities of matter.[55] These are shape, size, position, motion or rest, contiguity and number. (Toricelli, following up Kepler's work on gravity, later added 'mass', but Galileo did not consider 'mass' — or strictly speaking weight (*gravità*), since he did not have a concept of mass — to be a primary quality, and he places it somewhere between primary and secondary qualities.[56] Indeed, the fact that weight occupies this rather unsatisfactory position, which tends to undermine the whole distinction, may have been partly responsible for Galileo's later comparative silence on the question of primary and secondary qualities.) We may note that the primary qualities are either geometrical (shape, size, position, contiguity) arithmetical (number) or kinematic (motion or rest). Galileo makes a clear conceptual distinction between these primary qualities, which are absolute, objective and immutable, and the secondary qualities, which are relative, subjective, fluctuating and sensible. The realm of secondary qualities is that of illusion; the realm of primary qualities is that of knowledge, and it is this latter alone to which mathematics is applicable.[57] This thesis is the reverse of Aristotle's. For Aristotle, mathematics is necessarily restricted, since it is inapplicable to reality, which is not amenable to exhaustive quantitative analysis. On the doctrine of primary and secondary qualities, on the contrary, whatever is not mathematical 'in nature' does not really exist in nature at all; it is merely a psychic addition of the perceiving mind.

In the doctrine of primary and secondary qualities, we can find an epistemological basis on which the roles of mathematics and experiment can be articulated in a manner conducive to Galileo's Platonism. The doctrine could form the theoretical foundation for an elevation of the status of mathematics and a demotion of the status of experiment. But, contrary to received opinion,[58] the doctrine hardly figures after *Il Saggiatore*. Why is this, we may ask?

First, the doctrine of primary and secondary qualities is really no more than an intuitive abstraction from everyday experience. It does not simply duplicate naive sense perception (but then neither do the concepts of Peripatetic mechanics), but the visual and tactile senses retain their privileged epistemological status. The kind of mechanical concepts which could be formulated on this basis could not stray far from those of Peripatetic mechanics. Secondly, Galileo simply does not need, and does well to avoid, this doctrine. Abstraction from sense experience does not provide the source of concepts in later Galilean mechanics. In approaching the question of what *does* provide this source, we must turn to Galileo's reconceptualisation of the role of experiment.

The two main indices of Galileo's new conception of experiment are to be found in his use of the *Gedankenexperiment*, or thought experiment, and in his construction and use of scientific instruments. A kind of thought experiment — the *reductio ad absurdum* — had been used as early as Parmenides, and Aristotle had used this mode of argument to great effect throughout his work; for example, in the rejection of the existence of the void. The *reductio ad absurdum*, however, is completely negative and can only be used to reject certain possibilities. Galileo uses thought experiments in quite a different way. For example, the comparison of the speeds of two identical lead spheres when falling independently with their speeds when joined together is used not only to reject the Aristotelian doctrine of absolute weight, but also to introduce a theory of specific weight.

It is worth noting, moreover, that Galileo's apparently dismissive remarks about experiments cannot always be taken at face value. To escape from the uncertainty deriving from individual impressions in observation he constructed his experiments with the use of measuring techniques which were often quite new. He constructed the first instrument designed to measure temperature differences in two solids or liquids; his work on the weight of air, the speed of propagation of light, the strength of magnets, all involved novel measuring devices.[59] No-one could deny that there were serious short-comings in Galileo's experiments and in his instruments. We cannot use modern standards as a point of comparison, but Galileo's results do not compare well with Mersenne's value for the acceleration of free fall, for example.[60] But what is of central importance is not Galileo's ability as an experi-menter, but his conceptualisation of the role of experiments. These two aspects of Galileo's work have been confused only too often. The crucial point is that Galileo does not merely use his experiments to check his results.

Tycho Brahe, as is well known, was able to develop posi-tional astronomy to a remarkable degree of accuracy by using the quadrant and the astrolabe. But what is crucially different about Tycho's and Galileo's use of instruments is that the former uses them exclusively to support and refine the evidence of the senses. It was one of the basic tenets of the prevailing Aristotelian philosophy, within which Tycho worked, that true appearances are those given to perception under normal conditions. On this kind of account, scientific instruments are merely an extension of our perceptual facul-ties; they are not correctives, that is, things which can be used to question our normal perceptual faculties. Galileo's account is quite different from this. For him, the telescope, in particular, is 'a superior and better sense than natural and common sense' and much more able 'to join forces with reason'.[61] But not only do scientific instruments have a different status in Galileo's work, the role of experiment changes also. The rationally conceived experiment can now actually supply the basis on which a theory can be constructed.

We are now in a position to consider Simplicio's objection to laws concerning the fall of bodies in a void.

§6   Experiment as a Necessary Condition of Mathematisation

Simplicio's objection does not merely call a particular result into question, it calls the whole of Galileo's mode of proof and explanation in physics into question. What we have here, in its clearest and most perspicuous form, is the Aristotelian objection to the Galilean mode of proof. What is at stake is the question of what counts as an explanation in physics.

The objection is, then, not a trivial one. Indeed, we have already noted some of Galileo's failures to transform problems mathematically, and, in the case of diurnal motion, we have seen a geometrical representation completely obscure the real problem. Because of this, the way Galileo deals with the problem of the resistance of the medium in free fall requires an extended analysis. I hope to show in this analysis that the process of transforming a physical problem in such a way that it becomes anemable to mathematical treatment not only involves experimentation, but also that this experimentation is the basis on which the physical concepts, in terms of which the physical problem is to be posed, are formulated.

The laws which Galileo proposes with respect to his three main categories of motion — uniform motion, uniformly accelerated motion, and projectile motion — are supposed to have a general validity. This general validity can be established only if two factors, which affect the motion of bodies moving in non-vacuous media, can be dealt with adequately. They are the 'weight' of the body and the 'resistance' of the medium through which the body moves. However, the role and effect of these factors cannot be evaluated until their nature has been clarified. In Galileo's discussion of this problem in the 'First Day' of the *Discorsi*,[62] two main points are made. The first is that we must distinguish between a body's absolute weight and its specific weight; and that the former, as such, has no effect on a body's speed. The second point is that the resistance of a medium cannot be treated as a unitary phenomenon. There are in fact two distinct, though connected, kinds of resistance which a medium offers to motion, each having a different kind of effectivity.

Galileo's treatment of specific weight and resistance in the 'First Day' takes the form of a series of arguments, each of which is associated with an experiment or set of experiments. The use of experiment is remarkably rich; experiments are proposed in some cases as simple refutations of particular theories, but in others they function as a direct source of conceptual innovation.

The first set of arguments appears to be a direct refutation of the Peripatetic theory of the role of absolute weight and resistance in motion:

> '[Aristotle] makes two assumptions; one concerning moveables differing in heaviness but moving in the same medium, and the other concerning a given moveable moved in different mediums. As to the first, he assumes that moveables differing in heaviness are moved in the same medium with unequal speeds, which maintain to one another the same ratio as their weights [*gravità*]. Thus, for example, a moveable ten times as heavy as another, is moved ten times as fast. In the other supposition he takes it that the speeds of the same moveable through different mediums are in inverse ratio to the crassitudes or densities of the mediums. Assuming, for example, that the crassitude of water is ten times that of air, he would have it that the speed in air is ten times the speed in water.'[63]

This direct relation between speed and weight is rejected on two grounds. First, experience shows us that two bodies of the same material, but of very different weights, when dropped simultaneously from a height of two hundred braccia, arrive at the ground at the same time. As well as this *a posteriori* argument there is, secondly, an *a priori* one. If two bodies of differing weights, but again of the same material, fall independently then on the Aristotelian position we must say that they fall with different speeds. But, 'it is evident that were we to connect the slower to the faster, the latter would be partly retarded by the slower, and this would be partly speeded up by the latter'.[64] Hence the aggregate speed of the connected bodies would be less than the original speed of the heavier body, even though the aggregate weight would be greater than that of the heavier body. But this is a straightforward contradiction. Hence the speed of a falling body cannot be proportional to its absolute weight.

The conclusion is that 'both great and small bodies, of the same specific gravity, are moved with like speeds'.[65] The qualification 'of the same specific gravity' is a provisional one. The only bodies which have been compared in the preceding argument are ones having the same specific gravity. The precise role of specific gravity can only be evaluated once the role of resistance has been evaluated. Aristotle had argued that speed is inversely proportional to the resistance, or density, of the medium. Again, this theory is refuted by a simple *reductio ad absurdum*. If we say that the density (*corpulenza* — 'materiality') ratio of water to air is 10:1, and if we take a body which falls in air but floats in water, such as a wooden sphere, and say that this has a speed of fall of twenty units in air, then we must conclude that its speed of fall in water is two units. However, since it is one of our premises that the wooden sphere will float in water, it cannot also fall in water with a speed of two units. Hence speed cannot be inversely proportional to the density of the medium.

What has been established up to now is that speed is neither directly proportional to absolute weight nor inversely proportional to the resistance of the medium. The only positive conclusion is that speed is somehow proportional to specific weight. This is highly problematic, however.

In the *De Motu*, speed was calculated by subtracting the specific weight of the medium from that of the body. But this direct link between specific weight and speed of fall is incompatible with the principle that increase in speed is in a simple proportion to time, since we have no guarantee that the increase in speed will occur in exactly the same way for all bodies, irrespective of specific weight. This would mean that we could not demonstrate the identity of natural and uniform acceleration. A new relation between the specific weight of a falling body and the resistance of the medium is clearly required.

I said above that the first set of arguments *appeared* to be directed solely against the Aristotelian theory of speed. But in an important sense they are also directed against the *De Motu*. The experimental situations described in the *De Motu* correspond to those given in Aristotle and they are of a particular kind: speed of motion is determined with respect to the one body in two different media, or with respect to two different bodies in one and the same medium. In the *Discorsi*, this experimental situation is generalised to cover the motion of a body in any medium:

> 'I began to combine these two phenomena together, noting what happened with moveables of different heaviness placed in mediums of different resistances, and I have found that the inequality of speeds is always greater in more resistant mediums, as compared with those more yielding'.[66]

The initial experimental situation envisaged is one in which the speeds of two heavy metals are compared in two different media. Gold and lead fall with approximately the same speed in air, at least over the short distances which are amenable to observation on earth, but only the former sinks in mercury, the latter rising to the surface.[67] An important negative consequence here is that the experiment shows that the speed ratio of different bodies is not reflected directly in their specific weights. This is all the more instructive when we consider that the specific weights of bodies in air are approximately equal to their absolute specific weights, so we might reasonably expect that their specific weight ratio in air reflects the speed ratio. That this does not occur undermines the link between specific weight and speed.

This situation gives rise to two questions. First, since the differences in speed between any two bodies decreases with the density of the medium, what would happen in a medium with a density of zero? By extrapolation to a void, the conclusion Galileo offers is that 'in a void all speeds would be entirely equal'.[68] This extrapolation is justified if specific gravity and resistance are the only factors affecting speed in free fall.[69]

The second question is much more problematic: what exactly *is* the relation between speed, specific weight and resistance? In answering this question Galileo begins by putting forward the following experiment. Consider two bodies, ebony and lead, and two media, air and water. The ratios of their specific weights are as follows (we are interested in orders of magnitude rather than exact numerical values here):

	Medium	:	Body
Ebony in air	1	:	1,000
Ebony in water	800	:	1,000
Lead in air	1	:	10,000
Lead in water	800	:	10,000

It is clear that as the specific weight ratio approaches one, the speed of motion of the body tends towards zero. The difference between the specific weights in any one ratio is the resistance. This 'resistance' is in fact the buoyancy effect of the medium. Instead of speaking of 'resistance' we shall now distinguish the buoyancy effect, and this will be justified below.

The table shows that when ebony is transferred from air to water, the buoyancy effect increases in such a way that it now represents 0.8 of the body's specific weight, whereas in air it represented only 0.001 of its specific weight. In the case of lead, the increase is from 0.0001 to 0.08. That is to say, in air the buoyancy effect decreases the weight of lead and ebony by a factor of a thousandth and a ten thousandth respectively; both of these are negligible. In water, however, the ebony loses four fifths of its weight whereas the lead loses less than a tenth of its weight.

Because the buoyancy effect is so disproportionate for bodies of different specific weights it is clear that differences in specific weight *per se* cannot account for the difference in speed of fall of bodies. Specific weight now becomes only one of several factors which determine differences of speed in a given medium. There is a further conclusion to be drawn from this argument. We have noted that as the medium becomes rarer its buoyancy effect decreases. Let us now return to the question of speed in a void. It is clear that in a void there is no buoyancy effect since, as we have seen, the buoyancy effect is determined by the ratio between the specific weights of the body and the medium. But the differences in speed of bodies of different specific weights is also determined by this ratio. Since a void has no specific weight it cannot bear a ratio relation to the specific weight of the falling body. That is to say, this ratio, which determines differences in speed, cannot be operative in the case of a void, and hence we must conclude that all bodies — whatever their specific weight — fall with the same 'degrees of speed' in a void.[70]

Now this conclusion is particularly important since on the basis of the equality of speeds of all bodies in a void we can proceed, at least in principle, to calculate differences in speeds between any two bodies in any media by determining the amount by which the theoretical speed in a void will be diminished, where this diminution is calculated on the basis of the ratio of the specific weights of the medium and the freely falling body; the calculated difference in speed would then be relatively easy to verify experimentally. To this end, Galileo takes us through the (not very successful) experiments to measure the specific weight of air in a void.

There remains one outstanding problem: freely falling bodies accelerate. Simplicio poses the problem this way:

'If the difference in weight of moveables of different heaviness cannot cause the change [with distance] in the ratio of the speeds, because the heaviness does not change, then neither can the medium cause any alteration in the ratio of speeds, since it too is always assumed to stay the same.'[71]

In dealing with this problem, Galileo introduces into the discussion two elements of crucial importance. The first is:

'. . . that the heavy body has from nature an extrinsic principle of moving towards the common centre of heavy objects (that is, of our terrestrial globe) with a continually accelerated movement, and always equally accelerated, so that in equal times there are added new momenta and degrees of speed.'[72]

This is the meaning of the 'degrees of speed' referred to above; Galileo moves from the common idea that bodies have speeds in fall to the idea of their having uniformly accelerated motion in fall. From the earlier discussion, the positive conclusion about the equality of speeds of bodies in a void is now extended to cover changes of speed.

The second new element which Galileo introduces is a resistance effect of the medium which is distinct from its buoyancy effect:

'There is an increase of resistance in the medium, not because this changes in essence, but because of change in speed with which the medium must be opened and moved laterally to yield passage to the falling body that is successfully accelerated.'[73]

This new effect of the medium — which Galileo always treats, together with the buoyancy effect, under the generic heading of 'resistance' — can be called the *friction effect*.[74] The friction effect can account for the variation of the speed ratio of two bodies — and hence in the case of free fall, for the variation of the change of speed ratio — because, unlike specific weight and the buoyancy effect, it is a variable whose value is not fixed simply by specifying a given body and a given medium through which it falls. The variability of this effect arises from the fact that as a body accelerates the friction effect increases, since progressively larger amounts of the medium have to be traversed — and hence 'opened and moved laterally by the falling moveable' — per unit time. A state of equilibrium is reached when the speed ceases to increase and the body falls at constant speed. This state of equilibrium occurs much earlier in rarer bodies, not because the friction effect bears a direct relation to specific weight, but because the buoyancy effect is much greater in bodies of lower density, and hence their motion is already greatly retarded.

Now it is clear from this argument that in the absence of the friction and buoyancy effects — that is, in a void — a body would continue to accelerate at a uniform rate, and specific weight would have no effect. The conclusion is that:

'No difference in weight, however great, plays any part at all in diversifying the speeds of moveables, so that as far as speed depends on weight, all moveables are moved with equal celerity. At first glance, this seems so remote from probability that, if I did not have some way of elucidating it and making it clear as daylight, it would have been better to remain silent than to assert it. So now that it has escaped my lips, I must not neglect any experiment or reason that can corroborate it.'[75]

In fact, two kinds of experiment are needed. First, the friction effect has to be isolated in such a way that differences in speed can be ascribed to it alone. Several proposals are put forward for realising this circumstance as an experimental situation. Free fall from a great height is impracticable; also, the buoyancy and friction effects would not be distinguishable in such a situation. If, however, we could compare the simultaneous motions of two bodies of different weights in a succession of gradually increasing or decreasing distances of fall, where the distances are practicably small, 'the minimal differences of time that might intervene between the arrival of the heavy body at the terminus and that of the light one',[76] could be measured. Now in order to eliminate the buoyancy effect the motions must be made as slow as possible, and Galileo puts forward the suggestion that the bodies be moved down a plane of very low gradient. The problem of the plane's friction — that is, surface friction — arises here however. These difficulties are resolved by proposing an experiment in which two spheres — one of cork and one of lead, the latter being about a hundred times heavier than the former — are suspended on threads of equal length and set in oscillation. While the periods of oscillation remain identical for both spheres, the amplitude of swing is considerably reduced in the case of the cork sphere very quickly. Now this reduction cannot be due to the higher specific gravity of the lead conferring a greater capacity for speed on it, since if we begin the experiment by swinging the cork through a greater arc than the lead it will travel with a greater speed. Since the buoyancy effect bears a direct relation to the specific weight ratio, the reduction cannot be due to this either. Hence what is operative in this situation is the friction effect, and we can determine from the experiment that this is greater for rarer bodies.

It remains to be shown that the friction effect increases with speed. If the friction effect does increase with speed then if, because of an artificial *impeto*, a falling body moves with a speed greater than its maximum constant speed (for that medium), it will be retarded until this speed is reached. If the friction effect does not increase with speed then this will not happen. Now because of the great distances that would be involved, and the difficulties in measurement that would ensue, it would be almost impossible to deal directly with freely falling bodies. Hence the consequence that bodies will be retarded until they reach their natural maximum speed is of crucial importance, since the experimental situation in which this can be tested can be realised in a relatively easy and straightforward way. A gun is shot vertically downwards from a great height and the penetration of the bullet into the ground is measured. It is then shot nearer to the ground and the penetration is again measured. The first penetration is less than the second, which means that the bullet has been retarded.[77]

What has been achieved in the whole of this argument — from the initial rejection of the proportionality of absolute weight and speed to the analysis of the friction effect — is quite remarkable. It has been proved that differences in speed are not due to the differences in weight of falling bodies; rather, they are due to the resisting effect of the medium. This resisting effect has been carefully broken down and analysed, and a whole series of experiments has been put forward to establish each of the various arguments at every stage. The very exhaustive analysis of the resisting effect of the medium provides the justification for the theorem that all freely falling bodies undergo the same uniform natural acceleration.

§7 Ontological Problems: Idealisation and Reality

The free fall of bodies in a void is a situation which is amenable to rigorous mathematical analysis. It is a situation in which the free fall of all bodies is identical. By taking this as the primary case for analysis, one can then examine the specific factors which produce differences of speed for different bodies in different media and these can be determined in a no less rigorous fashion. That is to say, the nature of these factors — the extent to which they depend on one another and the extent to which they are independent — and their effects can be determined by a series of independent experiments. This means that the law of falling bodies cannot be rejected as a mathematical fiction of no physical import. The same considerations apply to the discussion of projectile motion in the 'Fourth Day' of the *Discorsi*.

Given that we are not dealing with mathematical fictions, are we nevertheless dealing with an 'idealisation' when we refer to the motion of bodies in a void? In answering this question we need to make some preliminary distinctions. Let us call a theorem or law which refers to an experimental situation which cannot be realised either in principle or in fact a 'strong' idealisation. Let us call a theorem or law which refers to an experimental situation which cannot be realised in fact, but which can be realised in principle, a 'weak' idealisation. Finally, let us say that a theorem or law which refers to an experimental situation which can be realised both in principle and in fact is not an idealisation at all. (An experimental situation is realisable in principle if there is no incoherency involved in conceptualising that experimental situation. For example, if one argues that a void is logically impossible then an experimental situation in which bodies move in a void is incoherent. If one argues that it is not logically impossible then no incoherency is involved.)

We may now ask which of these categories Galileo's treatment of bodies falling in a void comes under. It clearly comes under the second. Galileo explicitly rejects the Aristotelian denial of the logical possibility of there being a void.[78] This means that the experimental examination of bodies falling in a void is possible in principle if not in fact.[79] Hence we are dealing with a 'weak' idealisation here. Of the categories we have distinguished, that of a 'weak' idealisation is the most problematic. Let us see, therefore, how it bears up under a more detailed analysis.

There seem to be two kinds of 'weak' idealisation, or at least two distinct circumstances both of which come under the category 'weak idealisation'. In the first case, we have a circumstance in which the experimental situation cannot be realised in fact, and where there is no other experimental situation, or no other set of experimental situations, which could act as a substitute. This is a hypothetical case, and it seems unlikely that it could ever be said with certainty that there could be no substitute. The second case is not hypothetical. This is the case where we have a circumstance in which the experimental situation cannot be realised in fact, but where there are other experimental situations which can act as exhaustive substitutes. Now this second case is the one which occurs in Galileo's analyses of free fall and projectile motion. We cannot test the uniform acceleration of bodies falling freely in a void in a direct way. All experiments must be performed in a resisting medium. But if the experiments performed in a resisting medium can actually establish what would happen in a void then we no longer have any need of a direct experiment. The series of arguments and experiments to determine the precise nature and effect of the resistance of the medium, which we have discussed above, *are designed to do just this*. Hence, a situation arises where the law that all bodies fall with uniform acceleration in a void can be dealt with experimentally in a way that does not require that we actually perform the experiments in a void. But if this is the case, how does Galileo's account differ from the third category above, where we said that no idealisation was involved. The answer must be that it does not differ from the third category. No idealisation is involved.

This conclusion is an important one, for it leads us directly to some fundamental ontological issues. In order to present the problem of falling bodies (and, we may add, the problem of projectile motion) in a mathematically viable form, Galileo has to deal with a situation which no-one has had, or could have, direct experience of. But by treating the problem in the way he does — that is, by analysing the resistance effect of the medium — he establishes this situation as a *real* one. Something has happened to reality here; or, more strictly speaking, something has happened to the criteria by which something is established as being real. A situation which does not occur in reality is no longer equivalent to a situation of which we have, or could have, no direct experience.

This does not mean that the two realms which Aristotle distinguishes — the realm of forms and the realm of those things which have matter and form — become identical. More importantly, the subject matter of physics is not the same as that of mathematics. There is no disagreement on this between the Aristotelian and Galilean accounts. On the usual interpretation, this is where the disagreement comes, but I hope that we can now see why this is wrong. The disagreement comes on the question of how the subject matters differ and how they are related. What Galileo's analysis enables us to do is to transform the subject matter of physics in such a way that it becomes amenable to mathematical treatment. This transformation is effected by experiment.

Experimentation is not something which simply accompanies the mathematical treatment of physical problems, it is a necessary condition of the mathematisation of a physical problem. Because experiment has this role, the 'reality' which is dealt with in the mathematical treatment of a problem is not that of everyday experience: it is the 'reality' of carefully controlled physical experiments.

The manner in which the concepts of Galilean kinematics are formulated is quite distinct from the formation of concepts in Aristotelian physics. They are not simply the result of an abstraction from everyday experience, and in some cases principles are formulated which actually contradict this experience — as in the principle of the physical relativity of motion, for example. Physical concepts are formulated on the basis of experimentation, in such a way that these concepts can then be used in the posing of physical problems which are amenable to mathematical treatment, and hence to rigorous mathematical solution on the basis of proven mathematical theorems and tested mathematical techniques. Because these concepts are not simply abstracted from experience, the physical problems which are posed do not usually correspond to something which we have direct access to. The situation to which the physical problem refers is, nevertheless, not an idealisation: it is established logically and experimentally. Indeed, what is examined in Galileo's accounts of free fall and projectile motion is not primarily the motion of bodies as these are given in everyday experience, but the motions of bodies in a situation where all variables and parameters can be specified, and where these parameters and variables are specified by the state-variables which fix the domains of investigation and explanation of the discourse.

This does not mean that the perceived motions of bodies cannot be accounted for in Galilean mechanics; it does mean that the account which is given of these derives from an account of a quite different kind of situation which is specified in terms of state-variables which may be quite alien to everyday experience. The procedures involved here are not analogous to the astronomical procedure of explaining apparent motions in terms of real ones. An account like this does (very properly) figure in the description of bodies falling on a rotating earth, in the *Dialogo*. But the situation here is quite different to that of bodies falling free in a void. The account of free fall (or projectile motion) in a void does not render the motion of bodies in resisting media an *apparent* motion. If it did, Galileo would indeed be a Platonist of the first order, and Simplicio would have real cause for complaint.

This has an important bearing on how we take the reply which Salviati gives to Simplicio:

> 'No firm science can be given of such events of heaviness, speed, and shape, which are variable in infinitely many ways. Hence to deal with such matters scientifically, it is necessary to abstract from them. We must find and demonstrate conclusions abstracted from impediments, in order to make use of them in practice under those limitations that experience will teach us.'[80]

Galileo seems to be conceding Simplicio's objection here: for if we are merely dealing with abstractions, we are dealing with a situation which is not 'real', and this is the case even if the abstractions are justified on the grounds that anything other than abstractions are not amenable to rigorous treatment. But the statement cannot be taken this way. That is to say, we cannot treat it as an acknowledgement of the Aristotelian distinction between abstraction and reality. Were we to take it in such a way, the laws of falling bodies and projectile motion would be demolished: we would not have given an account of 'reality', and the laws would be second best to laws concerning 'reality'.

However, as we have seen, this treatment of 'reality', as something opposed to 'abstraction', is not in keeping with Galileo's accounts of either the law of falling bodies or the law of projectile motion. The situations which these laws describe are not 'unreal'. Nor are they 'special cases'; they are simply the primary cases to which Galileo directs his analysis. There are two reasons for this primacy. First, these cases are primary because they are the most general cases subject to rigorous mathematical treatment. That is to say, they are primary in mathematical physics. To argue that they are not primary 'in nature' is to fall back into Aristotelianism. Nothing is primary 'in nature'. The cases we treat as primary in our everyday experience have this status as a result of a conceptual structuring of everyday experience which may be, and Galileo shows *is*, wholly inappropriate when we come to deal with problems in physics.

Secondly, the primacy, for physics, of uniformly accelerated motion in a void derives not from its being the *simplest* case, but from its being *the most general case. Any* freely falling body is subject to uniform acceleration; in a resisting medium, however, certain other factors — which can be precisely specified and whose effects can be evaluated — act to determine the resultant motion of the body.

§8   The New Explanatory Structure

The manner in which the concepts of Galilean kinematics are formulated is quite distinct from the corresponding procedure in Aristotelian and Medieval physics. Concepts are not abstracted from sense experience, and in some cases principles are put forward which actually contradict this experience. This is one of the crucial differences between the Aristotelian and Galilean procedures in physics, but if taken in isolation from the concepts of evidence and proof which these physical discourses operate with, it is difficult to see what the rationale for the new procedure is. The difference arises primarily because of the different explanatory requirements of these discourses; in particular, it arises because of the difference in the way in which physical problems must be posed — and hence conceptualised — if explanations are to be generated in physics.

In both Aristotelian and Galilean physics, attempts are made to explain one kind of thing (or 'phenomenon' or 'occurrence') in terms of another kind of thing (or 'phenomenon' or 'occurrence'). The crucial difference lies in the kinds of things which figure in the explanations. In the case of the Aristotelian explanatory structure, it has been argued that the kinds of things which must figure in explanations (that is, ultimately, essences) cannot be linked to the domain of evidence specified in that structure, and this means that explanations cannot be generated. Galileo's radical reassessment of the issues involved in explanation in physics gives rise to a new situation, in which kinematic problems — conceptualised on an experimental basis — are transformed mathematically in such a way as to generate rigorous and verifiable explanations. Nevertheless, the results which are actually produced in Galileo's work are mainly kinematic results, and little — if any — progress is made in dynamics. The reasons for this are complex. One obvious problem is the lack of the appropriate mathematical concepts and techniques. This is not to say that there are no sections of dynamics which cannot be dealt with in geometrical terms. Newton's treatment of dynamics in the *Principia* operates largely in terms of geometrical demonstrations, but it is noteworthy that Newton is able to transcend geometry without too much difficulty, and he is able to appeal to calculus on the occasions when the need arises. Galileo has to rest content with geometry, and this is a serious hindrance in the presentation of dynamical problems.

Nevertheless, it would be absurd to suggest that the difficulties which Galileo experiences in dynamics are purely mathematical ones. The most serious problems, as might be expected, centre around the question of dynamical concepts. We have extolled Galileo's treatment of the buoyancy and friction effects above, but it must still be remembered that friction is treated simply as a hindrance to motion rather than a force, and that the dynamic problems involved in friction are just ignored. In the context, this has some justification, but at the more general level it represents a serious failing in Galilean physics. Moreover, we have noted that Galileo's attempts to deal with dynamical problems result, in the main, in a reduction of dynamics to kinematics or statics or even, in some cases, to geometry.

The question we must ask is: Does the explanatory structure that we find in Galileo's work provide an adequate basis for the mathematical presentation of dynamical problems? It may be noted that we are not asking whether this explanatory structure provides us with the requisite physical concepts. It cannot provide us with physical concepts, only with the procedures by which physical concepts can be generated: we are asking whether the appropriate kinds of concepts can be generated. It certainly cannot provide us with the appropriate mathematical concepts nor even with the procedures by which these can be generated, since it is a physical explanatory structure and not a mathematical one.[81] It can only circumscribe the kinds of mathematical concepts required, in terms of the functions which they must fulfil. Secondly, we have already noted that Galileo himself cannot pose dynamical problems adequately, so this is not at issue. What *is* at issue is whether dynamical problems can be posed adequately — and mathematically — on the *explanatory* basis provided in Galileo's work, and this problem is really one about whether the appropriate concepts can be formulated *without* this requiring a radical change in what counts as an explanation in physics. A full answer to this question would require a rather extensive treatment of classical mechanics, which would be outside the scope of this work, but some general points can be established, nevertheless, which indicate that the explanatory structure that we have discussed in the later parts of this chapter *does* provide a basis for classical physics.

In Galileo's later work on kinematics, physical concepts are formulated on the basis of experimentation in such a way that these concepts can then be used in the formulation of physical problems which are amenable to mathematical treatment. Because these concepts are not simply abstracted from everyday experience, the physical problems which are posed do not usually correspond to something that we have direct access to. The situation which the physical problem refers to is, nevertheless, not an idealisation. The entities treated in such a problem, and the relations between these entities, are established logically and experimentally. This ontological re-definition of the realm of physics is only inaugurated in Galileo's work on free fall and projectile motion. There are still Aristotelian elements which figure in the Galilean accounts — notably the idea of a bounded Cosmos, and the idea of 'natural place'[82] — and these prevent the full conceptualisation of such things as gravitational mass. Moreover, in Galileo's work we still find that bodies have many properties which are conferred on them intuitively rather than experimentally. Uniform rectilinear motion, for example, occurs only on a horizontal plane;

when the body leaves the plane its weight causes it to move downwards.

Nevertheless, the important point is that in certain crucial areas Galileo confers properties on bodies on a purely experimental basis. What is being sought in these cases are not *essences* but *parameters*. The statement of the law of falling bodies, for example, is not a statement of the essence of bodies. The series of experiments which lead up to the formulation of the law establish that the absolute and specific weights of bodies play no role in motion in a void. They do not establish that weight is not part of the essence of bodies because there are situations where differences in weight can be ignored. The procedure which is operative here is in stark contrast not only to the procedure followed by Aristotle, but also to that prescribed in Cartesian metaphysics and mechanics. At the level of explanation, in fact, Descartes' work is much further removed from classical mechanics than Galileo's. For Descartes, matter is a substance and motion is a mode of this substance. Matter and extension are identified on the basis of the following argument:

> 'But although any one attribute is sufficient to give us knowledge of a substance, there is always one principal property of substance which constitutes its nature and essence, and on which the others depend. Thus extension in length, breadth and depth, constitutes the nature of corporeal substance; and thought constitutes the nature of thinking substance. For all else that may be attributed to body presupposes extension, and is but a mode of this extended thing.'[83]

This argument is couched in strangely Scholastic terms. Matter is a substance and substance always has one and only one essence. This essence is extension. If we assume the substance/essence relation as given it remains to be determined why extension should be this essence. The answer is that we abstract from experience those and only those properties of matter which are such that if a body did not have them it would not be matter. In the wax example of the *Second Meditation*, for example, the only properties of wax which remain constant on melting are extension, flexibility and mobility. Extension is isolated as the only essential property here, on the grounds that all bodies are necessarily extended in that we cannot conceive of an unextended body.

In Cartesian metaphysics, although we actually perceive external objects — since God would not deceive us in this respect — our knowledge of external objects does not derive from 'the senses or from the faculty of imagination, but from our intellect alone'.[84] This argument works at two levels. First, although the world is given through sense perception, only those aspects of it which result from the process of abstraction — those which unchangingly remain — are the objects of proper understanding. Second, what remains after abstraction takes on an ontological significance: it is not just an abstraction, it is what really exists.

The procedure, then, is this. One abstracts from perceptual experience all dubitable qualities, and what remains is extension. What is interesting here is not so much the idea of a physics based on extension — which is not objectionable as such — but rather *how one arrives at* this physics. To make extension an ontologically primary notion because it remains as a residue from abstractions from sense experience is to suppose that perceived properties of bodies can form the basis for a mathematical physics. Now the postulate that all bodies are and must be extended is a commonplace in Aristotelian and Scholastic accounts of 'Nature' and it has some justification there, since not only can it be derived from the essential unity of matter and form, but also 'Nature' is taken as the totality of sensible things. To accept this postulate as a basis for a mathematical physics — as Descartes attempts to do — is an entirely different question.

The essential properties of bodies which Descartes is seeking find no place in the explanatory structure which we have described in this chapter. The properties sought are those which remain constant and relevant in a counterfactually specified situation which has general application and which is amenable to mathematical treatment. These properties are not those which any body of a particular kind must have in any possible world; rather, they are those properties which are operative in a certain specified situation. This situation has two particularly important features. First, it is *general*, inasmuch as the factors which are operative in it are also operative in all other relevant situations, that is all those cases which can be characterised in terms of the state-variables of the discourse. It is not subject to the variables which are peculiarly operative in these other situations, however, and this is what confers it generality on it. Secondly, the factors operative can be treated mathematically — that is, a quantitative account can be given and particular values can be checked experimentally.

Galileo's account does not simply introduce counterfactuals, it introduces counterfactuals of a particular kind — ones describing situations subject to full mathematical treatment. What is involved here is not just the mathematical formulation of physical problems, but also the construction of these problems in such a way that their solutions have general applicability. Avempace had operated with counterfactuals, and Oresme had treated physical problems in a mathematical fashion. What distinguishes Galileo's work is his ability to establish the generality of particular kinds of problem which he can then treat mathematically. In doing this he establishes experimentation as a necessary condition of posing physical problems mathematically. It is on the basis of a procedure of this kind — and not on the basis of a search for properties which bodies necessarily have — that explanations in terms of entities which far transcend what can be perceived, or what can be obtained from an abstraction from perception, become possible. In the same way that Galileo can deal with bodies whose actual weight — specific or absolute — is irrelevant, so later physicists could introduce mass points — where extension becomes irrelevant. No-one in the seventeenth century would have denied that all bodies have determinate extension and determinate weight. But these properties which all bodies have soon become irrelevant in the formulation of particular kinds of physical problems, and concepts and properties, which are based on neither intuition nor everyday experience, are introduced. This fundamental change occurs because of a change in the way physical concepts are developed and physical problems posed. In these last three chapters, I have tried to specify exactly what is involved in this change, the kinds of explanatory problems which required that such a change take place, and the nature of the issues which had to be dealt with if a workable account of explanation in physics was to be provided.

Notes: Chapter 6

1   We noted in ch. 4 that this geometrical mode of proof differs from algebra in that it is not properly symbolic insofar as although it works with a symbolism which is abbreviatory, this symbolism does not represent the workings of combinatory operations. Further, Greek geometry is strongly dependent on sensory experience. One of the clearest examples of this is the restriction conferred on multiplication, where no more than three lines can be multiplied. The reason for this is that the product of two lines was conceived as a plane, the product of three lines as a solid, and there are only three physical dimensions. These issues are discussed in Mahoney, 'Die Anfänge der algebraischen Denkweise', p.16 ff; and in Mahoney, *The Mathematical Career of Pierre Fermat*, p.41 ff. See also the literature cited above, ch.4, note 55. Galileo had no algebraic techniques at his disposal and this seriously restricted his mathematical treatment of physical problems: cf. Boyer, 'Galileo's Place in the History of Mathematics', and King, *Measurement and Natural Philosophy*, p.194 ff.

2   *Dialogo*, p.103.

3   *De Motu*, p.79.

4   Archimedes had, of course, also been concerned with the theory of lever-equilibria and problems of determining centres of gravity. These questions are dealt with in his 'On the Equilibrium of Planes' (Archimedes, *Works*, pp.189-220). His hydrostatics is presented in 'On Floating Bodies' (*ibid*, pp.253-300). As Dijksterhuis (*op. cit.*, p.53 ff) has stressed, both these are essentially mathematical rather than physical works. For example, the central proposition of Book II of 'On Floating Bodies' is Proposition 1: 'If a solid lighter than a fluid be at rest in it, the weight of the solid will be to that of the same volume of fluid as the immersed portion of the solid is to the whole' (p.263). That this result was conceived as being primarily mathematical, and not as a physical result, is clear from the way in which it is followed up in the subsequent propositions in Book II. No physical application of the principle is discussed, and instead we are immediately introduced to geometrical problems concerning the stability of segments of floating paraboloids of revolution. Similarly, in

dealing with lever-equilibria, Archimedes makes no attempt to apply his results to the theory of simple machines. Archimedes' project could, in fact, be seen in terms of Aristotle's account of the 'subordinate sciences'. When Aristotle speaks of 'optics' as being a subordinate science what is being referred to is the geometrical study of catoptrics and perspective, not the study of the nature of light. Similarly, statics — as practised by Archimedes — would fall under the general rubric of the subordinate sciences in that it deals mathematically with mathematical problems which, as a matter of fact, have physical application. Koertge ('Galileo and the Problem of Accidents', p.394 ff) has shown how Galileo, in the *De Motu* and *Le Meccaniche*, attempts to give a mathematical account of certain physical problems by simply imagining what would happen in the absence of factors such as the earth's curvature. What is instructive about these early attempts is that they work with 'idealisations' in the strict sense — Galileo can give no account of the conditions under which it is legitimate to ignore certain factors. This is exactly what is problematic about Archimedes' work construed as physics. The present chapter is primarily concerned with how this problem is solved in Galileo's later work.

5   Cf. *Physics*, 216a13 ff and Galileo's criticisms in his *De Motu*, pp.26-38.

6   The rigid separation of physical and geometrical bodies pervades Galileo's work up to and including the *Dialogo*. I shall discuss some of the issues involved in this separation below.

7   Westfall, *Force in Newton's Physics*, p.9.

8   Westfall (*ibid*, p.21) has argued that the postulated proportionality between specific weight and speed is incompatible with Galileo's 'atomism'. Galileo's gives a cautious blessing to the atomist thesis that there is 'one kind of matter in all bodies, and those bodies are heavier which enclose more particles of that matter in narrower space' (*De Motu*, p.14). That is to say, the density of a body is merely the result of the spacing of its 'elementary particles'. The apparent discrepancy arises when we consider that Galileo's rejection of the Aristotelian proportionality of absolute weight and speed is based on the argument that two identical lead spheres, connected by a thread, do not fall twice as quickly as they would if they were not connected; they fall at the same speed even though the absolute weight has doubled. But this argument is also applicable to that 'one kind of matter' which is the sole constituent of all things (including

rare media, presumably). Westfall concludes that 'the argument which established that all pieces of lead fall with the same speed should have demonstrated as well that all bodies whatsoever fall with the same speed'. This is a *non sequitur* however. What *is* established is that all bodies would fall with the same speed in a void (a conclusion that Galileo does not accept in the *De Motu*). In the case of fall through a resisting medium, on the other hand, it is in no way established that all bodies fall with the same speed. Consider two bodies: A has a density of four units, B has a density of two units. These fall through a resisting medium with a density of one unit. In atomist terms, this means that there are four atoms per unit volume of space in A, two in B, and one in the medium. This in turn means that, in free fall, for every four atoms of A there is a resistance of one atom of medium, and that for every two atoms of B there is a resistance of one atom of medium, and hence A will move through the medium faster than B. Further, two units of A connected together will meet the same resistance per unit volume of space as will one unit of A, so the lead spheres paradox will not arise. Qualitatively, if not quantitatively, this result is quite compatible with Galileo's account.

9   Cf. *De Motu*, pp.76-100.
10   *Discorsi*, p.173.
11   Cf. Westfall, *op. cit.*, p.25. A rather interesting case — which possibly goes against Westfall's general thesis that Galileo's dynamics is dominated by statics — is that in which the concept of 'moment of descent' [*momento di descendere*] figures (cf. *Le Meccaniche*, pp.169-177 and *Discorsi*, p.175 ff). The concept 'moment of descent' is to be distinguished from the other 'moments' that Galileo deals with, as these latter cannot be treated in isolation from a system of constraint, since their values can only be determined with respect to the point of the system to which the body is attached. Because the moment of descent on an inclined plane remains unchanged throughout the length of the plane it is not restricted to a point of a system. Further, by decreasing the angle of inclination of a plane we decrease the downward tendency of a body on that plane — its motor force — without changing its weight. This means that we must separate gravity *qua* motor force from gravity *qua* weight: the one concept *gravità* is not sufficient for the dynamical description of the motion of heavy bodies. The dynamical concept formulated in response to this situation is 'moment of descent'. However, since Galileo's analysis is

restricted to inclined planes and never generalised to cover free fall, the general importance of the concept — and its relation to the classical concept of momentum — is difficult to assess. A charitable assessment is given in Clavelin, *op. cit.*, pp.165-172, 346-359, and 482-3.

12 *Dialogo*, p.188 ff.
13 *Ibid*, p.125 ff.
14 *Ibid*, p.126 ff.
15 *Ibid*, p.132 ff.
16 *Ibid*, p.188.
17 I have followed Hall's simplified geometrical construction; cf. Hall, *Galileo to Newton*, p.52.
18 Boyer, 'Galileo's Place in the History of Mathematics', pp.246-7.
19 *Dialogo*, p.199.
20 *Ibid*, p.201. See Drake's discussion of this in his notes (*ibid*, p.478).
21 *Ibid*, p.197.
22 Cf. *Opere*, VIII, p.373.
23 Cf. Descartes, *Oeuvres*, X, p.219 ff. See also Koyré's now classic discussion of these two cases in his *Études Galiléennes*, pp.86-136. Unlike Galileo, Descartes never corrects the law. I think part of the reason for this is that motion and rest are modal contraries in Descartes' fully fledged metaphysics, and hence they are discontinuous. This means that a body starting to move from rest does not pass through all the degrees of speed (as Galileo maintained). This in turn precludes the geometrical representation of time and speed as straight lines meeting at a point where they have values of zero. What would result, in fact, would be a step function diagram.
24 *Dialogo*, p.213 ff.
25 *Ibid*, p.215.
26 *Ibid*, pp.215-6. Following Clavelin (*op. cit.*, p.242), I have added the constructions *GG'* and *EE'* to make the argument clearer.
27 *Dialogo*, p.217.
28 *Ibid*, p.217.
29 In this respect, Galileo's account is worth comparing with Huygens' 'classical' treatment of the problem. By exploring the geometrical properties of circles, Huygens is able to derive exact quantitative expressions for *vis centrifuga* in terms of the velocity of the body and the diameter of the circle in which it moves. For a clear exposition of Huygens' procedure see Westfall, *Force in Newton's Physics*, p.169 ff.

30 We shall see later that the situation in the *Discorsi* is rather different.

31 Cf. Dijksterhuis, *op. cit.*, pp.348-9.

32 *Dialogo*, p.19 ff and p.32.

33 *Ibid*, p.28. Galileo concludes not only that each planet was put into its orbit by being 'dropped' and then acted upon so as to have its rectilinear motion altered to a circular motion of the same speed, but also that all of the planets were so 'dropped' from one and the same place (cf. Koyré, *Newtonian Studies*, ch. 4; and Cohen, 'Galileo,. Newton and the Divine Order'). It may be noted that Galileo's commitment to 'order' does not mean that he believed that the planets actually move in 'perfect' circles with a uniform speed. He explicitly denies the uniformity of the motion of the earth around the sun, for example, in the 'Fourth Day' of the *Dialogo* (pp.453 and 455).

34 *Dialogo*, p.234.

35 Brown ('Galileo, the Elements and the Tides') has drawn attention to the importance of construing Galileo's statements in terms of 'earthy' as opposed to 'earthly' bodies. The former are those composed of the element 'earth', the latter are those which happen to be part of the planet Earth.

36 Coffa, 'Galileo's Concept of Inertia'. See also Drake, 'Semi-circular Fall in the *Dialogue*', and *Galileo Studies*, chs. 12 and 13.

37 This is pointed out in Wisan, 'The New Science of Motion', p.262 ff; see also Westfall, 'The Problem of Force in Galileo's Physics'.

38 *Discorsi*, p.221. This is a restatement of the Theorem which precedes the proof. The original statement reads: 'When a projectile is carried in motion compounded from equable horizontal and from naturally accelerated downward [motions], it describes a semiparabolic line in its movement' (*ibid*, p.217).

39 *Ibid*, pp.221-2.

40 *Ibid*, pp.222-3.

41 *Ibid*, p.223.

42 *Dialogo*, p.203.

43 Koertge, *op. cit.*, has offered a very interesting comparison of Guidobaldo and Galileo on this issue. She also presents a schematic account of Galileo's 'development' on the question, which she describes in terms of the different procedures for stripping away 'accidents'. I think this presentation of Galileo's work in terms of accidents is very misleading for two reasons. First, it suggests that what is left once the 'accidents'

have been stripped away is something which is essential and Galileo is certainly not looking for something essential in his later works. Secondly, even if we accept that the term 'accident' is not being used in this (its normal) sense, Koertge's account still suggests that what Galileo hopes to find once the 'accidents' have been stripped away remains constant. That is to say, it suggests that what change are the procedures by which the one project — common to all Galileo's work — is realised. My own view, as will be clear from this chapter, is that the project itself changes: in fact it changes so radically that to describe the procedures in the *Discorsi* in terms of stripping away accidents is to obscure the novelty of Galileo's achievements.

44 *Dialogo*, p.14. This is a late sixteenth century interpretation of Aristotle. It is not quite Aristotle's own position.

45 Cf. Galileo to Liceti, 15th September 1640; *Opere*, XVIII, pp.247-251 (esp. p.248).

46 Cf. McTighe, 'Galileo's Platonism'.

47 *Discorsi*, p.245.

48 Cf. Koyré, *Metaphysics and Measurement*, chs. 2 and 5; and Feyerabend, *Against Method*, *passim*. Koyré and (to a lesser extent) Feyerabend are generally very critical of Galileo's use of experiment. This general view has been challenged — both implicitly and explicitly — by a number of authors. See in particular: Drake, 'Galileo's Experimental Confirmation of Horizontal Inertia', Geymonat, *Galileo Galilei*, p.174 ff; Moscovici, *L'Expérience du Mouvement*, p.140 ff; Schmitt, 'Experience and Experiment'; Settle, 'Galileo's Use of Experiment'.

49 Cf. *Discorsi*, p.169 ff. It is perhaps worth noting that the determination of this value is not an integral part of Galileo's project. His formula for free fall is $S_1:S_2 = T_1{}^2:T_2{}^2$ and not $s = \frac{1}{2}gt^2$, and given this the determination of g is not necessarily a particularly important problem for him. Cf. Settle, *Galilean Science*, pp.4-5.

50 Koyré, *Metaphysics and Measurement,* p.94.

51 *Discorsi*, p.170; for a detailed discussion of this and connected experiments see Settle, *Galilean Science*, ch. 2. Settle shows that the experiments to verify $S_1:S_2 = T_1{}^2:T_2{}^2$ are in fact quite accurate.

52 Koyré, *Metaphysics and Measurement*, p.38.

53 Burtt, *The Metaphysical Foundations of Modern Physical Science*, p.64 ff.

54 Cf. Strong, *Procedures and Metaphysics*, Chs. 2 and 8.

55 *Il Saggiatore*, p.274 ff. Cf. also Burtt, *op. cit.*, p.45 ff.

56 Cf. Koyré, *Études Galiléennes*, p.145 ff. The word *'mole'*, which is sometimes rendered as 'mass', is in fact more correctly translated as size or bulk. Cf. Shea, *Galileo's Intellectual Revolution*, p.45.

57 In this connection see his attack on Sarsi's claim that perception alone can decide between physical theories (*Il Saggiatore*, p.255).

58 E.g. in Burtt, *op. cit.*, p.73ff; Hall, 'The Significance of Galileo's Thought', p.73; Namer, 'L'Intelligibilité Mathématique', p.119.

59 Cf. Bedini, 'The Instruments of Galileo Galilei'; also Clavelin, *op. cit.*, p.402. Clavelin considers that Galileo fitted an elementary micrometer to his telescope in 1612. This is surely wrong. The only telescope that he had access to in 1612 was the 'Galilean' telescope, which has a concave eyepiece and a convex object lens. The *real* image on such a telescope is located at infinity, and the *observed* image is seen at the near point — which depends, amongst other things, on the focal length of the eye. In such a situation a micrometer cannot be used for comparing images. This is a serious failing of the Galilean telescope since at a time when the magnification of lenses was difficult to determine, a knowledge of the comparative sizes of images for any one telescope was crucial if results were to be compared.

60 Cf. Koyré, *Metaphysics and Measurement*, p.99 ff.

61 *Discorsi*, p.328.

62 *Ibid*, pp.65-108. In my account of this part of the 'First Day' I have made considerable use of Clavelin's detailed commentary (*op. cit.*, ch. 7).

63 *Discorsi*, p.65. Cf. Aristotle, *Physics*, 215a24 ff.

64 *Discorsi*, p.67.

65 *Ibid*, p.68.

66 *Ibid*, p.72.

67 *Ibid*, p.75.

68 *Ibid*, p.76.

69 Shape can be treated as a function of resistance.

70 *Ibid*, p.77. Drake, in his commentary, notes that up to this point Galileo has dealt with free fall in terms of fixed natural speeds, for simplicity of treatment. The introduction of 'degrees of speed', in the plural, opens the way for the subsequent discussion of acceleration.

71 *Ibid*, p.77.

72 *Ibid,* p.77.

73 *Ibid*, p.78.

74 Galileo does not *explicitly* distinguish between the buoyancy and friction effects, but the conceptual, if not the verbal distinction between these two is adhered to throughout the *Discorsi*. As Clavelin (*op. cit.*, p.336) has noted, the only occasion on which the two are conflated, insofar as the buoyancy effect is neglected in a case in which it is operative, is in the discussion of a cannonball moving through two media (*Discorsi*, p.95).

75 *Discorsi*, p.86.

76 *Ibid*, p.87.

77 *Ibid*, p.95. See Westfall's comments on this kind of procedure in his 'The Problem of Force in Galileo's Physics', p.83.

78 Cf. *Discorsi*, p.65 ff.

79 So far as I know, this is still strictly true for bodies larger than molecules since every such body at a temperature above absolute zero is subject to vaporization, and if vaporization occurs the body can no longer be said to be in a vacuum.

80 *Discorsi*, p.225.

81 I must admit that I am not yet sure whether mathematical discourses have explanatory structures at all. This is a problem that I hope to deal with in a subsequent publication.

82 Galileo's conception of 'natural place' is rather different from Aristotle's. Cf. Koyré, *Études Galiléennes*, p.240.

83 Descartes, *Oeuvres*, IX_2, p.48. (Principles, I, 53.)

84 *Ibid*, VII, p.34. (Second Meditation.)

CONCLUSION

CHAPTER 7

THE ANALYSIS OF EXPLANATORY STRUCTURES

§1 Summary of the Main Arguments

IT has long been recognised — at least implicitly — that theoretical discourses, at different stages of their development, incur explanatory failures of one kind or another. What has not always been realised is that the kinds of explanatory failure that discourses are subject to — and the kinds of explanatory achievement that they make possible — are analytically inseparable from the peculiar constraints that are operative on explanation in different discourses. It is clear that there must be constraints on explanation in discourses, otherwise anything would count as an explanation and any one account would be as good as any other. I have been concerned primarily with identifying these constraints and with examining the way in which they vary, and in particular with the reasons for this variation and the way in which it allows certain kinds of explanatory problem to be overcome.

Before we consider some of the general issues that are raised by this project it may, perhaps, be worthwhile summarising the main points that I have tried to establish. In Part I, I attempted to construct a theory of explanatory structures by examining the concepts of ontology, proof and evidence, and by trying to specify the general roles which ontologies, proof structures and domains of evidence play in explanation. With regard to ontology, I have argued that the ontology of a discourse consists of a set of entities bearing definite but complex relations to one another. Great stress has been put on the fact that these entities are related in a definite fashion, for ontologies may consist of the same entities related in different ways, and these different relations have a crucial bearing on the kinds of account which must be given in proposing explanations. For example, Classical Atomism and Cartesian metaphysics (and their respective physical theories) both operate with ontologies in which matter and space figure, but because of the different ways in which these are related the kinds of explanations which are being sought ultimately differ. These issues are central to the account that I have proposed because it is primarily in terms of its ontology that explanations in a discourse are given. The ontology of a discourse determines the kinds of entities which can figure in explanations: it also determines what relations these entities can bear to one another. In this sense it plays a crucial role in circumscribing the kinds of accounts that are candidates for explanations.

I have introduced the idea of a domain of evidence to designate that set of phenomena which could support, confirm or tend to falsify possible explanations given in a particular discourse. To be schematic, we can say that whereas the ontology of a discourse is a crucial factor in delimiting criteria of appropriateness for explanation, the domain of evidence, in determining what counts as evidence in a discourse, is a crucial factor in delimiting the criteria of adequacy for explanations. Further, since in proposing explanations in terms of an ontology reference must be made to a domain of evidence, there must be procedures for relating the accounts given in explanations to evidential information. I have argued that this connection is effected by a system of concepts which links the situation which is accounted for in the explanation to situations which have evidential value and to which we have access. I have also argued that the rules governing the inferences which are allowable from statements operating in terms of this system of concepts can be conceived in terms of the proof structure of a discourse, and that differences in proof structures — and particularly, in the specific cases that we have considered, differences between syllogistic and geometrical proof structures — involve not only differences in the way problems are solved but also in the way problems are posed. It is this latter issue that I have concentrated on, and I have attempted to connect the form which explanations must take in a discourse to the way in which problems are to be posed in that discourse.

In Part I, problems of ontology, evidence and proof have been discussed and, on the basis of this discussion, an account of explanatory structures has been proposed. In Part II, Aristotelian and Galilean physics have been examined in terms of their explanatory structures. The aim of Part II has been to determine what counts as an explanation in these discourses, whether explanations of the required kind can be given and, if they cannot be given, why this is the case and how the situation can be remedied (if at all).

In the case of the explanatory structure of Aristotelian physics, it has been argued that explanations of the kind required in this physics cannot be given in principle. There are several reasons for this, the main one being that what is sought in these explanations cannot be linked, conceptually, to the domain of evidence of Aristotelian physics. A complementary problem, which arises much more clearly in the Middle Ages and in the Renaissance, is the lack of criteria by which to decide when an account offered is of the required explanatory kind. I have argued that these problems are inherent in the nature of the explanatory structure of Aristotelian and Medieval mechanics. They cannot be overcome by simply changing some part of this structure. In particular, the problems are such that the attempt to introduce mathematical proofs into physics, while at the same time seeking explanations of the Aristotelian kind, cannot be successful. I have tried to show that such an attempt is not even well-considered, since mathematical proofs are precluded from Aristotelian physics on a highly articulate and well thought-out basis. This argument has been supported by a detailed analysis of the mode of proof with which some Medieval theories — primarily kinematic theories — operate, and this analysis has revealed that what are usually taken to be rudimentary mathematical accounts are in fact closer to 'linguistic' analyses.

In the case of Galilean physics, I have attempted to show that the mode of formulation and resolution of physical problems that we find in this physics, and particularly in the *Two New Sciences*, constitutes a basic reappraisal of the issues involved in physical explanation. Problems of ontology, proof and evidence are all involved in this re-appraisal. I have argued that Galileo's work, far from being *a priorist*, employs experimentation as a means whereby problems can be posed mathematically. Experiment becomes a source of conceptual innovation; it does not function merely as a procedure for checking results. Philosophical accounts of experimentation usually restrict themselves to this latter function and I have argued that this is seriously misleading. Indeed, I have tried to establish that the primary function of experimentation in (later) Galilean mechanics is to provide a basis on which physical problems can be con-ceptualised in such a way that they become amenable to mathematical treatment, and that this is one of the main things that serves to distinguish Galilean mechanics from Aristotelian and Medieval mechanics, where experiment is used (if at all) *solely* to confirm results. Moreover, by relat-ing the problem of the function(s) of experiment to the problem of how one develops concepts which allow problems to be posed in specific ways — *viz*, in ways which yield solu-tions which have explanatory power — I have tried to connect these questions to more general issues concerning explanation. It has, in fact, been part of my overall argument that questions of evidence, concept formation and explana-tion are intimately linked and that each imposes constraints on the others: the actual form and power of these constraints varying depending on the discourse in question and on the kinds of problems which that discourse must face at different stages of its development.

The explanatory structure of Galilean mechanics is far from being problem-free, but it does provide a viable basis on which explanatory problems in mechanics can be dealt with. The basic explanatory failures to which the explanatory structure of Aristotelian mechanics is prone are such that this structure can only be abolished: revisions, such as attempts to alter the proof structure so as to allow the use of mathe-matical operations in physics, are inappropriate. In Galileo's work, on the other hand, fundamental developments and revisions are required, but these are needed in order that a viable account of explanation can be realised in dynamics (and ultimately in non-mechanical areas of physics such as electricity and magnetism), not because explanations of the proposed kind cannot be given in principle.

I do not wish to claim that this difference in explanatory requirements and achievements is the only difference between pre-classical and classical mechanics, or that other differences are ultimately reducible to this one. Nevertheless, in analysing this explanatory difference in detail, what has come to light is that there is a fundamental gulf between pre-classical and classical mechanics: at the level of explanation, the one is neither a development of, nor a revised version of, the other.

§2   General Issues

I argued in chapter 1 that there is no 'natural' differentiation of discourses: it is inappropriate to consider that some differentiations correspond more exactly to 'the way things are' than others. This is not to deny that some differentiations are confused or uninformative. Nor is it to deny that there are some issues — such as the question of whether theories have an overlap in reference in their domains of investigation — which do not depend on how we differentiate discourses. But there is nothing 'natural' about a classification of discourses in terms of their domains of investigation, and a classification in terms of explanatory structures, for example, would not include everything with the same referential domain of investigation in the same discourse, nor would most methodological or epistemological classifications. Indeed, when one considers the attempts to treat all theories whose domains of investigation refer to the same phenomena as being part of the same discourse, it soon become clear that this kind of characterisation is liable to be misleading, in that it is usually implicitly premissed on the assumption that different theories, in 'physics' for example, simply conceptualise the same 'physical phenomena' in different ways. But one decides whether something is a physical phenomenon or not on the basis of such a conceptualisation. Physical phenomena do not carry labels stating 'I am a physical phenomenon'. A physical phenomenon is such if any only if it is the referent of a concept specifying what 'physical phenomena' are, and concepts of what 'physical phenomena' are vary from discourse to discourse. The variation in these concepts occurs primarily at the level of sense, and because sense determines reference, on the only decent theory of meaning that we have (at present), it *may* also involve a change in reference. What we must be careful to avoid is the idea that concepts with different senses automatically have different references. The Fregean account of sense and reference entails no such thing, and indeed one of the standard Fregean examples, that of the morning star and the evening star, is precisely an example of two concepts with different senses but the same reference. I am not claiming that the issues in the development of physics are as straight-forward as this — far from it — but the general principle nevertheless holds good. What justifies (or rather what *would* justify) our calling Aristotelian, Newtonian and Quantum physics 'physics' in the first place is the fact (if it is a fact) that the references of the concepts which specify what appears in the domains of investigation have some overlap.

With regard to the conventional nature of how discourses are characterised, it is clear that there is nothing to prevent a classification on which Aristotelian and classical mechanics turn out to be part of the one discourse. On other classifications, what I have called 'Aristotelian mechanics' may turn out to comprise several distinct discourses. Such charac-terisations may be the result of attempts to pose quite different kinds of questions and they may not necessarily be comparable, although this would not rule out our being able to criticise the projects of which these characterisations of discourse are part, on the grounds that they cannot provide answers to the questions which they are designed to answer for example. In this case, we could argue that some characterisations are the products of ill-conceived projects.

Bearing this in mind, let me now turn to the characterisa-tion of discourses in terms of explanatory structures. I have argued that there is a fundamental gulf between pre-classical and classical mechanics. In speaking of a 'gulf' I do not mean a discontinuity *simpliciter* but a very specific type of discontinuity: namely, one which involves a fundamental change in what counts as an explanation in physics. The identification and analysis of a change of this kind is clearly important, but it is worthwhile asking why this requires the introduction of the idea of an explanatory structure, and where the strength of the analysis of explanatory structures lies.

The answer to the first question is, I think, relatively straightforward. In order to be able to give a full account of changes in explanatory requirements we must have some means of identifying what counts as an explanation in that discourse. This is not an easy matter. In the cases that we have been concerned with, for example, if we were simply to take the general philosophical and methodological statements of the practitioners of these discourses at their face value we would be in a very weak position. These statements are rarely consistent and they often consist of platitudes and propaganda (as Feyerabend has repeatedly remarked); similarly, the precedents which are often invoked in the work of earlier thinkers — particularly Greek philosophers and the Christian theologians — are usually quite anachronistic and inappropriate, and they are often designed merely to present new theories to a conservative audience in the guise of relatively familiar and unobjectionable traditional accounts. This is not to claim that these statements can be ignored, only that the discrepancy between, say, the procedures that a scientist actually operates with and those which he recommends in explicit philosophical theories of method is often a very real one.

A related problem, which perhaps deserves mention here, concerns the case in which theories from one discourse are taken up and developed in another. In some cases theories developed within a discourse with an unworkable explanatory structure — such as the Merton account of motion — may be taken up and developed in a quite different discourse which operates with a viable explanatory structure — such as classical mechanics. The important point here is that the criteria of successfulness or pertinence differ depending on the kind of physical discourse one is working within. The reasons why Bradwardine's account of the dynamic conditions of motion, for example, marks an advance in Peripatetic physics is that it resolves a fundamental discrepancy in Aristotle's own work. The way in which this account is to be assessed in classical mechanics is quite different. Particular medieval theories assessed in their own terms may mark a significant advance. These theories assessed in terms of classical mechanics may mark no advance at all. The criteria in both cases may differ vastly. Thus what is of interest is why Galileo, for instance, took up some particular Medieval theories and not others which may have commanded equal or even greater respect from Medieval physicists. That is, why were these theories open to fruitful development and not others? The answer to this question is to be found in an examination of *classical mechanics and not Medieval mechanics*. The Merton account of non-uniform motion may have some interest in its own right, but it has a quite different kind of interest *vis-à-vis* classical mechanics. It is only by examining the explanatory structures of the discourse in which the theory is first developed and the discourse in which it is subsequently taken up, developed and revised, that we can give this issue any detailed systematic treatment because it is only then that we can ask what the Merton School requires out of their accounts of motion and what Galileo requires out of these.

This brings us to the second issue. There are two questions in particular which, I think, the analysis of explanatory structures is especially helpful in elucidating. The first concerns the elaboration of new concepts and theories in a discourse. In characterising explanatory structures in terms of ontologies, domains of evidence, proof structures and systems of concepts I have tried to avoid the idea that discourses can be treated purely in terms of some central set of concepts or some central set of laws. The trouble with this latter kind of treatment is that there is a tendency to treat discourses in a wholly synchronic fashion. This leads to problems when we come to consider the question of why discourses are open to certain kinds of developments and not others. It is true that the system of concepts which a discourse operates with at any one time will constrain the subsequent development of that discourse, but this in itself tells us nothing about how new concepts are formulated and old ones discarded. It is for this reason that the characterisations of discourses in terms of systems of concepts (or laws) is very limited, and if we are to be able to give a diachronic account of discourses it does seem that we must also include, in our characterisation, the procedures by which these concepts (or laws) can be revised or discarded, and the procedures by which new concepts (or laws) can be generated. The characterisation of explanatory structures that I have proposed is designed to enable us to give some account, not only of the systems of concepts with which discourses operate, but also of the procedures by which concepts are generated in discourses.

The other main issue that I have tried to elucidate by introducing explanatory structures is the problem of how we deal with the role of logical and mathematical concepts in physical theories. I have argued that there is an intimate connection between the system of concepts with which a discourse operates and its proof structure. In chapter 3, this connection was examined at a general level, and in Part II specific cases were examined. In particular, I have tried, in the three main cases, to articulate the relation between the theory of the demonstrative syllogism and Aristotelian mechanics; *logica moderna* (particularly supposition theory and the theory of syncategorematic terms) and Merton kinematics; and classical Euclidean geometry and Galilean kinematics. Separate accounts of these three topics exist — and I have made use of these, as far as possible, in my own accounts — but there is no general treatment of the relations between proof structures and systems of concepts, and it is such a general treatment that I have tried to provide. Part of the motivation for raising this issue at a detailed level has been to show that the use of mathematics in physics requires justification, that such justification cannot be given in terms of 'idealisations', and that certain conditions must be fulfilled before physical problems can be posed mathematically in classical mechanics.

In establishing these specific points I have also tried to raise some issues of more general importance, if only at an elementary level. In particular, I have suggested that proof structures are not simply 'applied' to the concepts of a discourse, at least not in any straightforward sense. Mathematics, for example, is not simply 'applied' to classical mechanics in such a way that both mathematics and physics remain unchanged as a result of the application; or, to concentrate on the physical rather than the mathematical issues, a mathematical physics cannot be reduced to a non-mathematical physics plus mathematics. Aristotelian physics is a non-mathematical physics and it could not be otherwise; classical physics is a mathematical physics and it could not be otherwise. Just as syllogistic logic is an intimate part of Aristotelian mechanics, so too is ('classical') mathematics an intimate part of classical mechanics. The systems of concepts with which these discourses operate are *systems* inasmuch as the concepts are related in a definite fashion (since they serve to relate a definite ontology to a definite domain of evidence). What it means for concepts to be related in a definite fashion is that certain modes of proof, inference etc. are possible between these concepts, or at least between the propositions employing these concepts. The constraints on these modes of proof and inference are determined by the proof structure of the discourse, whether this be syllogistic theory, classical mathematics or whatever. It may be noted, however, that the relations between a proof structure and a system of concepts are *reciprocal*. Changes in the conceptual system of a discourse may lead to a change in its proof structure; conversely, changes in the proof structure may lead to revision in the system of concepts.

In the two explanatory structures that we have considered in detail, the main shift was from a proof structure which operated primarily in terms of the theory of the syllogism (together with the supplements and revisions made by the Medieval 'modernists') to a proof structure which operated primarily in terms of Euclidean geometry. This 'shift' was not independent of questions of explanation, evidence and ontology, as we have seen. The importance of linking changes in proof structure to these questions extends far beyond the specific cases that we have examined: indeed, in the currently problematic area of quantum mechanics a shift in proof structure has often been envisaged which, if it proves successful, may well require a change in our ideas about what counts as an explanation in micro-physics which is at least as radical as that inaugurated by Galileo in classical mechanics. All the proof structures that we have considered in this book have, logically speaking, been classical. All syllogistic and Euclidean operations obey the law of the excluded middle, for example, but there is no *a priori* reason why *any* proof structure should be classical. The use of non-classical geometries in Relativity Theory is generally accepted and, as we have noted in chapter 3, the use of non-classical logics has been suggested in quantum mechanics. Such a use has a pragmatic justification: if we allow non-distributive operations, for example, many anomalies are cleared up. In seeking more convincing reasons for allowing (or disallowing) non-classical logics, and non-standard mathematical analyses, in quantum mechanics it may be of some help to look at the explanatory structure of quantum mechanics (for the sake of the illustration we can assume that this constitutes the one clearly-defined discourse in explanatory terms). In this case what would be problematic would not be simply why quantum mechanics may only be able to operate properly with a non-classical logic and mathematics, but also why classical mechanics can operate properly only with classical logic and mathematics. The general kind of problem involved here is, I suggest, essentially no different from that involved when we consider the use of syllogisms and classical algebra in Aristotelian and classical mechanics respectively. Non-classical logics are often very counter-intuitive, and they may well have no use in everyday discourse, but this is no objection to their use in a discourse whose ontology admits of no intuitive grasp and which has a domain of evidence where, because of the Indeterminacy Principle, severe problems of measurement and calculation (at least by classical criteria) arise.

This brings us to a final issue. Although the account of explanatory structures that I have proposed is designed to have a general applicability — it should, in fact, cover *any* theoretical discourse — the only detailed analysis I have made are those contained in this book. Hence, although my general account is not completely dependent on these case studies, there can be no doubt that it is informed by the kinds of problems that I have come up against in the development of early mechanics. Many of these problems are specific to early mechanics, and these are doubtless many explanatory problems in other areas which an examination of early mechanics will be of little help in dealing with. This is, however, exactly what we would expect since one of the main conclusions that I have tried to establish in this book is that while, at the general level, we can specify those factors which play a part in determining what counts as an explanation, the way in which these factors operate in any particular discourse cannot be determined in advance. They bear a complex relation to one another which varies depending on the kinds of project that are set in a discourse in particular cases, and on the kinds of problems that are encountered in particular cases. The theory of explanatory structures that has been proposed is designed to provide us with a basis which allows the detailed investigations that are required in the various situations that arise to be carried out in a reasonably rigorous and comprehensive manner. The theory has been put forward to enable us to deal with the very different kinds of explanatory problems which arise in different discourses: the aim has *not* been to find out what is a 'good explanation' *per se*. I would not hold up Galileo's procedures in the *Discorsi*, for example, as a general paradigm of explanation in *any* area because there is simply no reason to suppose that these procedures can be extrapolated to other discourses where different ontologies, conceptual systems, domains of evidence and proof structures may be operative.

Finally, the tools that I have tried to elaborate in Part I, and use in Part II, are crude. They are open to development, revision and (so long as they are replaced by better ones) rejection. This is, I think, the only basis on which the theory that I have proposed can be genuinely informative.

BIBLIOGRAPHY

This bibliography lists only the works mentioned in the footnotes; it is not a comprehensive list of the literature on early mechanics.

ABELSON, P., *The Seven Liberal Arts*, New York, 1906.

ACKRILL, J. L., *Aristotle's Categories and De Interpretatione*, Oxford, 1963.

ACKRILL, J. L., 'Aristotle's Distinction between Energeia and Kinesis', in R. Bambrough (ed), *New Essays in Plato and Aristotle*, London, 1965.

ALBRITTON, R., 'Forms of Particular Substances in Aristotle's Metaphysics', *Journal of Philosophy*, LIV (1957), pp.699-708.

ANSCOMBE, G. E. M., 'Causality and Determination', in E. Sosa (ed), *Causation and Conditionals*, London, 1975.

AQUINAS, T., *Commentary on Boethius' De Trinitate*, trans. R. E. Brennan, London, 1946.

ARCHIMEDES, *The Works of Archimedes*, ed. and trans. T. L. Heath, Cambridge, 1897.

ARISTOTLE, *Aristotelis Opera*, ed. I. Bekker, Oxford, 1837, 11 volumes.

ARISTOTLE, *The Works of Aristotle Translated into English*, ed. W. D. Ross and J. A. Smith, Oxford, 1908-1952, 12 volumes.

AUBENQUE, P., *Le Problème de l'Être chez Aristote*, Paris, 1972.

AUSTIN, J. L., *Sense and Sensibilia*, London, 1964.

BACHELARD, G., *Les Intuitions Atomistiques,* Paris, 1935.

BACHELARD, G., *La Formation de l'Esprit Scientifique*, Paris, 1938.

BARNES, J., *Aristotle's Posterior Analytics*, Oxford, 1975.

BEDINI, S. A., 'The Instruments of Galileo Galilei', in E. McMullin (ed), *Galileo*, New York, 1967.

BIRKENMAJER, A., *Études d'Histoire des Sciences en Pologna,* Wrocław, 1972.

BOEHNER, P., *Medieval Logic*, Manchester, 1952.

BOYER, C., 'Galileo's Place in the History of Mathematics', in E. McMullin (ed), *Galileo*, New York, 1968.

BOYER, C., *A History of Mathematics*, New York, 1968.

BRADWARDINE, T., *Tractatus de Proportionibus*, ed. and trans. H. L. Crosby, Madison, 1955.

BRAMPTON, C. K., 'Scotus, Ockham and the Theory of Intuitive Cognition', *Antonianum*, XL (1965), pp.449-466.

BRIDGMAN, P. W., *The Logic of Modern Physics*, New York, 1958.

BROWN, H. I., 'Galileo, the Elements and the Tides', *Studies in the History and Philosophy of Science*, VII (1976), pp.337-351.

BUCHDAHL, G., *Induction and Necessity in the Philosophy of Aristotle*, London, 1963.

251

BURTT, E. A., *The Metaphysical Foundations of Modern Physical Science*, London, 1932.

ČAPEK, M., *The Philosophical Impact of Contemporary Physics*, Princeton, 1961.

CARNAP, R., *Foundations of Logic and Mathematics*, Chicago, 1939.

CARTERON, H., *La Notion de Force dans le Système d'Aristote*, Paris, 1923.

CASSIRER, E., *Das Erkenntnisproblem in der Philosophie und Wissenshaft der Neuren Zeit*, Berlin, 1922-3, 3 volumes.

CHARLTON, W., *Aristotle's Physics Books I and II*, Oxford, 1970.

CLAGETT, M., 'Richard Swineshead and Late Medieval Physics', *Osiris*, IX (1950), pp.131-161.

CLAGETT, M., *The Science of Mechanics in the Middle Ages*, Madison, 1959.

CLAGETT, M., *Nicole Oresme and the Medieval Geometry of Qualities and Motions*, Madison, 1968.

CLAVELIN, M., *The Natural Philosophy of Galileo*, Cambridge (Mass.), 1974.

COFFA, J. A., 'Galileo's Concept of Inertia', *Physis*, X (1968), pp.261-281.

COHEN, I. B., 'Galileo, Newton and the Divine Order of the Solar System', in E. McMullin (ed), *Galileo*, New York, 1967.

COLISH, M. L., *The Mirror of Language*, New Haven, 1968.

CROMBIE, A. C., 'Quantification in Medieval Physics', *Isis*, LII (1961), pp.143-160.

CROMBIE, A. C., *Robert Grosseteste and the Origins of Experimental Science*, Oxford, 1971.

CROMBIE, A. C., 'Sources of Galileo's Early Natural Philosophy', in M. L. R. Bonelli and W. R. Shea (eds), *Reason, Experiment and Mysticism in the Scientific Revolution*, London, 1975.

CUMMINS, R., 'States, Causes and the Law of Inertia', *Philosophical Studies*, XXIX (1976), pp.21-36.

DAUMAS, M., *Les Instruments Scientifiques aux XVIIe et XVIIIe Siècles*, Paris, 1953.

DAY, S. J., *Intuitive Cognition*, St Bonaventure, 1947.

DESCARTES, R., *Oeuvres de Descartes*, ed. C. Adam and P. Tannery, Paris, 1897-1910, 12 volumes.

DIJKSTERHUIS, E. J., *The Mechanisation of the World Picture*, London, 1961.

DRAKE, S., 'Semicircular Fall in the *Dialogue*', *Physis*, X (1968), pp.89-100.

DRAKE, S., *Galileo Studies,* Ann Arbour, 1970.

DRAKE, S., 'The Uniform Motion Equivalent to a Uniformly Accelerated Motion from Rest', *Isis*, LXIII (1972), pp.28-38.

DRAKE, S., 'Galileo's Experimental Confirmation of Horizontal Inertia: Unpublished Manuscripts', *Isis*, LXIV (1973), pp.291-305.

DRAKE, S. and DRABKIN, I. E. (eds), *Mechanics in Sixteenth Century Italy*, Madison, 1969.

DUHEM, P., *Études sur Léonard de Vinci*, Paris, 1905-1913, 3 volumes.

DUHEM, P., *Le Système du Monde*, Paris, 1913-1959, 10 volumes.

DUMMETT, M., *Frege: Philosophy of Language*, London, 1973.

EDDINGTON, A., *The Nature of the Physical World*, London, 1935.

ELLIS, B., 'The Origin and Nature of Newton's Laws of Motion', in R. Colodny (ed), *Beyond the Edge of Certainty*, New Jersey, 1965.

ERICKSON, C., *The Medieval Vision*, New York, 1976.

EUCLID, *The Thirteen Books of Euclid's Elements*, ed. and trans. T. L. Heath, New York, 1956, 3 volumes.

FEYERABEND, P. K., 'Explanation, Reduction and Empiricism', in H. Feigl and G. Maxwell (eds), *Minnesota Studies in the Philosophy of Science*, Vol. 3, Minneapolis, 1962.

FEYERABEND, P. K., 'Realism and Instrumentalism', in M. Bunge (ed), *The Critical Approach to Science and Philosophy*, Glencoe, 1964.

FEYERABEND, P. K., 'Problems of Empiricism', in R. Colodny (ed), *Beyond the Edge of Certainty*, New Jersey, 1965.

FEYERABEND, P. K., 'Zahar on Einstein', *British Journal for the Philosophy of Science*, XXV (1974), pp.25-28.

FEYERABEND, P. K., *Against Method*, London, 1975.

FINKELSTEIN, D., 'Matter, Space and Logic', in R. S. Cohen and M. W. Wartofsky (eds), *Boston Studies in the Philosophy of Science*, Vol. 5, Dordrecht, 1969.

FITZGERALD, J. J., 'Matter in Nature and Knowledge of Nature', in E. McMullin (ed), *The Concept of Matter*, Notre Dame (Ind.), 1963.

GALILEO, *Le Opere di Galileo Galilei*, ed. A. Favaro, Florence, 1968, 20 volumes.

GALILEO, *De Motu* [*c1590*], in I. E. Drabkin and S. Drake (eds and trans), *On Motion and On Mechanics*, Madison, 1960.

GALILEO, *Le Meccaniche* [c1600], in I. E. Drabkin and S. Drake, *On Motion and On Mechanics*, Madison, 1960.

GALILEO, *The Assayer* [*Il Saggiatore*, 1623], trans. and abr. in S. Drake (ed), *Discoveries and Opinions of Galileo*, New York, 1957.

GALILEO, *Dialogue Concerning the Two Chief World Systems* [*Dialogo*, 1632], trans. S. Drake, Berkeley, 1953.

GALILEO, *Two New Sciences* [*Discorsi*, 1638], trans. S. Drake, Madison, 1974.

GAUKROGER, S. W., 'Bachelard and the Problem of Epistemological Analysis', *Studies in the History and Philosophy of Science*, VII (1976), pp.189-244.

GEYMONAT, L., *Galileo Galilei*, New York, 1965.

GILBERT, N. W., *Renaissance Concepts of Method*, New York, 1960.

GILSON, E., *The Christian Philosophy of St Augustine*, London, 1961.

GILSON, E., *The Christian Philosophy of St Thomas Aquinas*, London, 1961.

GRANT, E., 'The Arguments of Nicholas of Autrecourt for the Existence of Interparticulate Vacua', *Actes du XIIe Congrès International d'Historie des Sciences, Paris 1968*, Vol.3A, Paris, 1971.

HAACK, S., *Deviant Logic*, London, 1974.

HALL, A. R., *From Galileo to Newton*, London, 1970.

HALL, A. R., 'The Significance of Galileo's Thought for the History of Science', in E. McMullin (ed), *Galileo*, New York, 1967.

HAMLYN, D. W., *Sensation and Perception*, London, 1961.

HAMLYN, D. W., *Aristotle's De Anima Books I and II*, Oxford, 1968.

HESSE, M. B., *The Structure of Scientific Inference*, London, 1974.

HINTIKKA, J., 'The Semantics of Modal Notions and the Indeterminacy of Ontology', in D. Davidson and G. Harman (eds), *Semantics of Natural Language*, Dordrecht, 1972.

HINTIKKA, J., *Time and Necessity*, Oxford, 1973.

HINTIKKA, J. and REMES, U., *The Method of Analysis*, Dordrecht, 1974.

HOOYKAAS, R., *Humanisme, Science et Réforme*, Leyden, 1958.

JAEGER, W., *Aristotle*, Oxford, 1962.

JAMMER, M., *Concepts of Space*, Cambridge (Mass.), 1969.

JAMMER, M., *The Philosophy of Quantum Mechanics*, New York, 1974.

JARDINE, L., *Francis Bacon*, London, 1974.

JARDINE, N., 'Galileo's Road to Truth and the Demonstrative Regress', *Studies in the History and Philosophy of Science*, VII (1976), pp.277-318.

KING, W. J. *Measurement and Natural Philosophy*, unpublished Ph.D. dissertation, Cornell University, 1961.

KIRK, G. S. and RAVEN, J. E., *The Presocratic Philosophers*, London, 1963.

KLEIN, J., *Greek Mathematical Thought and the Origin of Algebra*, Cambridge (Mass.), 1968.

KNEALE, W. and M., *The Development of Logic*, London, 1962.

KOERTGE, N., 'Galileo and the Problem of Accidents', *Journal of the History of Ideas*, XXXVIII (1977), pp.389-408.

KOSMAN, L. A., *The Aristotelian Backgrounds of Bacon's Novum Organon*, unpublished Ph.D. dissertation, Harvard University, 1964.

KOSMAN, L. A., 'Aristotle's Definition of Motion', *Phronesis*, XIV (1969), pp.40-62.

KOSMAN, L. A., 'Understanding, Insight and Explanation in the *Posterior Analytics'*, in E. N. Lee, A. D. P. Mourelatos and R. M. Rorty (eds), *Exegesis and Argument*, Assen, 1973.

KOYRÉ, A., *Newtonian Studies*, London, 1965.

KOYRÉ, A., *Études Galiléennes*, Paris, 1966.

KOYRÉ, A., *Metaphysics and Measurement*, London, 1968.

KOYRÉ, A., *Études d'Histoire de la Pensée Philosophique*, Paris, 1971.

KOYRÉ, A., *Études d'Histoire de la Pensée Scientifique*, Paris, 1973.

KRETZMANN, N., *William of Sherwood's Introduction to Logic*, Minneapolis, 1966.

KRETZMANN, N., 'Semantics, History of', in P. Edwards (ed), *The Encyclopaedia of Philosophy*, Vol. 7, New York, 1967, pp.358-406.

KRETZMANN, N., *William of Sherwood's Treatise on Syncategorematic Words*, Minneapolis, 1968.

KRETZMANN, N., 'Incipit/Desinit', in P. K. Machamer and R. G. Turnbull (eds), *Motion and Time, Space and Matter*, Ohio, 1976.

KUHN, T. S., *The Structure of Scientific Revolutions*, Chicago, 1970.

KUHN, T. S., 'Reflections on my Critics', in I. Lakatos and A. Musgrave (eds), *Criticism and the Growth of Knowledge*, London, 1972.

KUHN, T. S., 'Second Thoughts on Paradigms', in F. Suppe (ed), *The Structure of Scientific Theories*, Urbana, 1974.

LAKATOS, I., 'History of Science and its Rational Reconstructions', in R. C. Buck and R. S. Cohen (eds), *Boston Studies in the Philosophy of Science*, Vol. 8, Dordrecht, 1971.

LAKATOS, I., 'Falsification and the Methodology of Scientific Research Programmes', in I. Lakatos and A. Musgrave (eds), *Criticism and the Growth of Knowledge*, London, 1972.

LAKATOS, I. and ZAHAR, E. G., 'Why did Copernicus supersede Ptolemy?', in R. S. Westman (ed), *The Copernican Achievement*, Berkeley, 1975.

LEE, H. D. P., 'Geometrical Method and Aristotle's Account of First Principles', *Classical Quarterly*, XXIX (1935), pp.113-124.

LEFF, G., *William of Ockham*, Manchester, 1975.

LESCHER, J. H., 'The Meaning of *Nous* in the *Posterior Analytics*', *Phronesis*, XVIII (1973), pp.44-68.

LINDBERG, D. C., *Theories of Vision from al-Kindi to Kepler*, London, 1976.

LINDBERG, D. C. and STENECK, W. H., 'The Sense of Vision and the Origins of Modern Science', in A. G. Debus (ed), *Science, Medicine and Society in the Renaissance,* Vol. 1, London, 1972.

ŁUKASIEWICZ, J., *Aristotle's Syllogistic*, London, 1957.

MACH, E., *The Science of Mechanics*, La Salle (Ill), 1960.

MAHONEY, M. S., 'Another Look at Greek Geometrical Analysis', *Archive for the History of the Exact Sciences*, V (1968), pp.318-348.

MAHONEY, M. S., Die Anfänge der algebraischen Denkweise im 17. Jahrhundert', *Rete*, I (1971), pp.15-31.

MAHONEY, M. S., 'Babylonian Algebra: Form Vs Content', *Studies in the History and Philosophy of Science*, I (1971), pp.369-80.

MAHONEY, M. S., *The Mathematical Career of Pierre de Fermat*, Princeton, 1973.

MAIER, A., *Die Vorläufer Galileis im 14. Jahrhundert*, Rome, 1949.

MAIER, A., *Das Problem der intensiven Grösse in der Scholastik*, Rome, 1951.

MAIER, A., *Zwei Grundprobleme der scholastischen Naturphilosophie*, Rome, 1951.

MAIER, A., *An der Grenze von Scholastik und Naturwissenschaft*, Rome, 1952.

MAIER, A., *Zwischen Philosophie und Mechanik*, Rome, 1958.

MAXWELL, G., 'The Ontological Status of Theoretical Entities', in H. Feigl and G. Maxwell (eds), *Minnesota Studies in the Philosophy of Science*, Vol. 3, Minneapolis, 1962.

McTIGHE, T. P., 'Galileo's Platonism: A Reconsideration', in E. McMullin (ed), *Galileo*, New York, 1967.

MELLOR, D. H., 'Physics and Furniture', in N. Rescher (ed), *Studies in the Philosophy of Science*, Oxford, 1969.

MOODY, E. A., *Truth and Consequence in Medieval Logic*, Amsterdam, 1953.

MOODY, E. A., *Studies in Medieval Philosophy, Science and Logic*, Berkeley, 1975.

MOSCOVICI, S., *L'Éxpérience du Movement*, Paris, 1967.

NAGEL, E., *The Structure of Science*, London, 1961.

NAMER, E., 'L'Intelligibilité Mathématique et l'Expérience chez Galilée', in S. Delorme (ed), *Galilée, Aspects de sa Vie et de son Oeuvre*, Paris, 1968.

NYE, M. J., *Molecular Reality*, London, 1972.

ONG, W. J. *Ramus, Method and the Decay of Dialogue*, Cambridge (Mass.), 1958.

OWEN, G. E. L., 'Tithenai ta Phainomena', in J. M. E. Moravcsik (ed), *Aristotle*, London, 1968.

OWENS, J., *The Doctrine of Being in the Aristotelian Metaphysics*, Toronto, 1963.

OWENS, J., 'Matter and Predication in Aristotle', in E. McMullin (ed), *The Concept of Matter*, Notre Dame (Ind.), 1963.

PAETOW, L. J., *The Arts Course in Medieval Universities*, Urbana, 1910.

PEDERSON, O. and PIHL, M., *Early Physics and Astronomy*, London, 1974.

PETER OF SPAIN, *Summulae Logicales*, ed. and trans. J. P. Mullally, Notre Dame (Ind.), 1945.

PINBORG, J., *Die Entwicklung der Sprachtheorie im Mittelalter*, Münster/Copenhagen, 1967.

PINBORG, J., *Logik und Semantik im Mittelalter*, Stuttgart/Bad Cannstatt, 1972.

PLATO, *The Dialogues of Plato*, trans. B. Jowett, Oxford, 1953, 4 volumes.

POPPER, K. R., *The Logic of Scientific Discovery*, London, 1968.

PRESTIGE, G. L., *God in Patristic Thought*, London, 1936.

PUTNAM, H., *Philosophical Papers*, London, 1975, 2 volumes.

QUINE, W. V. O., *Word and Object*, Cambridge (Mass.), 1960.

QUINE, W. V. O., *Set Theory and its Logic*, Cambridge (Mass.), 1969.

RANDALL, J. H., *The School of Padua and the Emergence of Modern Science*, Padua, 1961.

RANDALL, J. H., *Aristotle*, New York, 1962.

REIF, P., 'The Textbook Tradition in Natural Philosophy, 1600-1650', *Journal of the History of Ideas*, XXX (1969), pp.17-32.

ROBINS, R. H., *Ancient and Medieval Grammatical Theory in Europe*, London, 1951.

ROBINSON, R., *Plato's Earlier Dialectic*, London, 1953.

RONCHI, V., 'Complexities, Advances and Misconceptions in the Development of the Science of Vision', in A. C. Crombie (ed), *Scientific Change*, London, 1963.

RONCHI, V., 'Galilée et l'Astronomie', in S. Delorme (ed), *Galilée, Aspects de sa Vie et de son Oeuvre*, Paris, 1968.

RONCHI, V., *The Nature of Light*, London, 1970.

ROSENTHAL, F., *Knowledge Triumphant*, Leyden, 1970.

Ross, W. D., *Aristotle's Prior and Posterior Analytics*, Oxford, 1949.

Ross, W. D., *Aristotle's Metaphysics*, Oxford, 1953. 2 volumes.

Sandys, J. E., *A History of Classical Scholarship*, Cambridge, 1921, 3 volumes.

Schmitt, C. B., 'Experimental Evidence for and against a Void: The Sixteenth Century Arguments', *Isis* LVIII (1967), pp.352-366.

Schmitt, C. B., 'Experience and Experiment: A Comparison of Zabarella's View with that of Galileo in the *De Motu*', *Studies in the Renaissance*, XVI (1969), pp.80-138.

Scott, D., 'Advice on Modal Logic', in K. Lambert (ed), *Philosophical Problems in Logic*, Dordrecht, 1970.

Sellars, W., 'The Language of Theories', in H. Feigl and G. Maxwell (eds), *Current Issues in the Philosophy of Science*, New York, 1961.

Sellars, W., *Science, Perception and Reality*, London, 1963.

Sellars, W., 'Scientific Realism or Irenic Instrumentalism', in R. S. Cohen and M. W. Wartofsky (eds), *Boston Studies in the Philosophy of Science*, Vol. 2, New York, 1965.

Settle, T. B., *Galilean Science*, unpublished Ph.D. dissertation, Cornell University, 1966.

Settle, T. B., 'Galileo's Use of Experiment as a Tool of Investigation', in E. McMullin (ed), *Galileo*, New York, 1967.

Shapere, D., *Galileo*, Chicago, 1974.

Sharp, D. E., *Franciscan Philosophy at Oxford in the Thirteenth Century*, Oxford, 1930.

Shea, W. R., *Galileo's Intellectual Revolution*, London, 1972.

Sheldon-Williams, I. P., 'The Greek Christian Platonist Tradition from the Cappadocians to Maximus and Eurigena', in A. H. Armstrong (ed), *The Cambridge History of Later Greek and Early Medieval Philosophy*, London, 1970.

Skulsky, H., 'Paduan Epistemology and the Doctrine of One Mind', *Journal for the History of Philosophy*, VI (1968), pp.341-361.

Solmsen, F., *Aristotle's System of the Physical World*, New York, 1960.

Steneck, N. H., *The Problem of Internal Senses in the Fourteenth Century*, unpublished Ph.D. dissertation, University of Wisconsin, 1970.

Straker, S. M., *Kepler's Optics*, unpublished Ph.D. dissertation, Indiana University, 1970.

Strawson, P. F., *Individuals*, London, 1959.

Strong, E. W., *Procedures and Metaphysics,* Berkeley, *1936.*

Sylla, E., 'Medieval Quantifications of Qualities: the Merton School', *Archive for the History of the Exact Sciences*, VIII (1971), pp.9-39.

Szabó, A., *Anfänge der grieschischen Mathematik*, Munich, 1969.

Unguru, S., 'On the Need to Rewrite the History of Greek Mathematics', *Archive for the History of the Exact Sciences*, XV (1975), pp.67-114.

Wallace, W. A., 'The Enigma of Domingo da Soto', *Isis*, LIX (1968), pp.384-401.

Wallerand, G., *Les Oeuvres de Siger de Courtrai*, Louvain, 1913.

Wang, H., 'On Denumerable Bases of Formal Systems', in T. A. Skolem et al., *Mathematical Interpretation of Formal Systems*, Amsterdam, 1955.

WEISHEIPL, J. A., 'Matter in Fourteenth Century Science', in E. McMullin (ed), *The Concept of Matter*, Notre Dame (Ind.), 1963.

WESTFALL, R. S., 'The Problem of Force in Galileo's Physics', in C. Golino (ed), *Galileo Reappraised*, Berkeley, 1966.

WESTFALL, R. S., *Force in Newton's Physics*, London, 1971.

WIELAND, W., *Die Aristotelische Physik*, Göttingen, 1970.

WILSON, C., *William Heytesbury*, Maddison, 1956.

WISAN, W. L., 'The New Science of Motion: A Study of Galileo's *De Motu Locali*', *Archive for the History of the Exact Sciences*, XIII (1974), pp.103-306.

WOODS, M. J., 'Problems in *Metaphysics* Z, Chapter 13', in J. M. E. Moravcsik (ed) *Aristotle*, London, 1968.

ZAHAR, E. G., 'Why did Einstein's Programme supersede Lorentz's?', *British Journal for the Philosophy of Science*, XXIV (1973), pp.95-123 and 223-262.

ZILSEL, E., 'The Genesis of the Concept of Physical Law', *Philosophical Review*, LI (1942), pp.245-279.

INDEX

abstraction, 88, 98ff, 123, 142ff, 161ff, 202, 222, 225ff
actuality and potentiality, Aristotle on, 114ff
Aegidius Romanus, 154
Albert of Saxony, 175n
Albertus Magnus, 140, 178n
Alhazen, 164, 179n
Anaximander, 114
Anselm, 136-7, 138
Aquinas, T., 135, 137ff, 145-6, 150, 154, 172n, 175n, 178n
Archimedes, 165, 184, 185-6, 197, 229-30n
Aristotle, 10, 18, 22, 23-32, 34n, 70, 71, 73, Ch's 4 and 5 *passim*, 186, 192, 226, 241ff
atomism, 34n, 42ff, 56ff, 92, 105, 230-1n
Augustine, 135-8, 168
Aurifaber, J., 173n
Austin, J.L., 50ff
Avempace, 150, 160ff, 228
Averroes, 140, 150, 176n
Avicenna, 140, 164

Bachelard, G., 5, 59n, 60n
Bacon, R., 164, 179n
Barnes, J., 95
being, order of and order of knowing, 89, 94ff, 134, 140ff
belief; Aristotle on, 112; Plato on, 109ff
Benedetti, G. V., 163ff
Bernard of Arezzo, 145
Boethius, 137
Bohr, N., 44
Bonamico, F., 185
Bradwardine, T., 148ff, 159ff, 176n, 178n, 245-6
Brahe, T., 209
Bridgman, P. W., 58n
Buridan, J., 160ff, 175n, 185
Burtt, E. A., 203

calculation, 74-5, 102
Callipus, 131

Čapek, M., 56, 58n, 60n
Carnap, R., 58n
Carteron, H., 133n
Cassirer, E., 161, 203
cause; Aquinas on, 140ff; Aristotle on (*aition*), 87ff; Plato on, 111ff
Clavelin, M., 155, 232n, 235n
Coffa, J. A., 196ff
Colish, M., 134
concepts; production of, 69-77, 164-6, 221ff, 246ff; systems of, 15, 67-9, 240
Copernicus, N., 9-11
counterfactuals, 227ff
Crombie, A. C., 177n
Cummins, R., 26-9, 35n

deductive-nomological model of explanation, 14
definition, Aristotle on, 87, 89
Democritus, 105, 122
demonstration, *see* syllogism, demonstrative
Descartes, R., 10, 43, 193, 226ff, 232n
Diophantus, 102
Domingo de Soto, 158
Donatus, 172n
Duhem, P., 45
dynamics; Bradwardine on, 148, 159ff; Buridan on, 162ff; Galileo on, 184ff, 189ff, 224ff, 231-2n

Eddington, A., 41
Einstein, E., 9-11
elements; Aristotle on, 120ff; Plato on, 110
Ellis, B., 25, 27ff
empiricism, 161
essence, 20ff, 34n, 226; Aristotle on, 84ff, 103ff, 107, 122ff, 126n; Descartes on, 226ff; Ockham on, 144ff; Putnam on, 20ff

259

proof structure, 15, 72-77, 240;
see also, mathematics and
physics; syllogism
Pseudo-Aristotle, 133n, 174n
Ptolemy, 9-11, 190
Putnam, H., 19ff, 74, 79n

Quine, W. V. O., 57-8n

Ramus, P., 170
Randall, J. H., 90, 95, 180-1n
reduction, 40, 76
reference, stability of, 17ff, 244
regressus theory, 166ff
relativity; optical, 119; physical,
48, 119
replacement, theoretical, 51ff
resistance; Aristotle on, 159,
177n, 186; Bradwardine on,
159ff; Galileo on, 186, 210ff;
Pseudo-Aristotle on, 174n
Ronchi, V., 179n

Scheckius, J., 167
Scott, D., 59-60n
Scotus, D., 142, 149, 154, 178n
Sellars, W., 46, 52
sense perception, 40ff, 50, 147;
Aquinas on, 138ff; Aristotle
on, 90, 91ff, 98ff, 105, 119ff,
122ff, 202; Avempace on,
161; Benedetti on, 165;
Descartes on, 226ff; Galileo
on, 209; Plato on, 109ff; and
reality, 40ff, 201, 207ff
Shapere, D., 153, 158
Siger of Brabant, 178n
Siger of Courtrai, 173n
Signa Dei, doctrine of, 137ff
signification, *see* language
Simplicius, 154
Sizzi, F., 179-80n
Skulsky, H., 168, 181n
species, 139, 142
state-variables, 22ff, 175-6n
statics, 167, 184ff, 193, 229-30n
statistical-relevance model of
explanation, 14

Strawson, P. F., 58-9n
'subordinate' sciences, 103ff,
147, 230n
substance; Aristotle on, 86ff,
127n; Descartes on, 226ff
supposition, *see* language
Swineshead, R. ('The
Calculator'), 155
syllogism, 75, 83ff, 248;
conjectural, 168;
demonstrative, 83ff, 102ff,
123, 166, 202; dialectical, 83,
92; eristic, 83; middle terms
of, *see* principle; premisses
of, 83-4, 104
Szabó, A., 129n

Tartaglia, N., 166
Theodoric of Freiberg, 164
theoretical discourses; and
atheoretical discourses, 32,
50; differentiation of, 3-13,
243ff; explanatory structure
of, 14-16, 29ff, 77-8, 239ff
thought experiments, 208ff
Torricelli, E., 207

universals; Aquinas on, 140ff;
Aristotle on, 86ff;
commensurate universals,
Aristotle on, 85ff, 97ff, 202;
Ockham on, 142; Zabarella
on, 169
unmoved mover, Aristotle on,
131n

void; Aristotle on, 120ff;
Avempace on, 160ff; Galileo
on, 67, 201ff, 218ff

Westfall, R. S., 230-1n, 231-2n
Wieland, W., 87, 131n
William of Sherwood, 137
Wilson, C., 158
Witelo, 179n

Zabarella, J., 167ff, 180-1n
Zahar, E., 9-11
Zeno, 118-132n